How We Hurt

How We Hurt

The Politics of Pain in the Opioid Epidemic

MELINA SHERMAN

OXFORD
UNIVERSITY PRESS

Oxford University Press is a department of the University of Oxford. It furthers the University's objective of excellence in research, scholarship, and education by publishing worldwide. Oxford is a registered trade mark of Oxford University Press in the UK and certain other countries.

Published in the United States of America by Oxford University Press
198 Madison Avenue, New York, NY 10016, United States of America.

© Oxford University Press 2024

All rights reserved. No part of this publication may be reproduced, stored in a retrieval system, or transmitted, in any form or by any means, without the prior permission in writing of Oxford University Press, or as expressly permitted by law, by license, or under terms agreed with the appropriate reproduction rights organization. Inquiries concerning reproduction outside the scope of the above should be sent to the Rights Department, Oxford University Press, at the address above.

You must not circulate this work in any other form
and you must impose this same condition on any acquirer.

Library of Congress Cataloging-in-Publication Data
Names: Sherman, Melina, author.
Title: How we hurt : the politics of pain in the opioid epidemic / Melina Sherman.
Description: New York, NY : Oxford University Press, [2024] |
Includes bibliographical references and index. |
Identifiers: LCCN 2023011323 (print) | LCCN 2023011324 (ebook) |
ISBN 9780197698235 (paperback) | ISBN 9780197698228 (hardback) |
ISBN 9780197698259 (epub) | ISBN 9780197698266
Subjects: LCSH: Opioid abuse—United States. | Chronic
pain—Treatment—United States.
Classification: LCC HV5825 .S4499 2024 (print) | LCC HV5825 (ebook) |
DDC 362.29/3—dc23/eng/20230322
LC record available at https://lccn.loc.gov/2023011323
LC ebook record available at https://lccn.loc.gov/2023011324

DOI: 10.1093/oso/9780197698228.001.0001

Paperback printed by Marquis Book Printing, Canada
Hardback printed by Bridgeport National Bindery, Inc., United States of America

To all of those who live in pain.

Contents

Acknowledgments   ix

Opening   1
1. Tracing the Painkiller Revolution   20
2. Strategic Ignorance in Opioid Regulation   46
3. Branding Pain Relief   75
4. Self-Help and the Rise of the Pain Patient-Expert   102
5. Pain's New Faces   118
Closing   147

References   161
Index   185

Acknowledgments

I did not write this book alone. Many people, in many different places around the world, helped me do it.

The first to acknowledge are the following members of my dissertation committee in Los Angeles, who read and reread this text in its earliest form and told me that it was, in fact, a book.

Patti Riley, thank you for helping me figure out how to write a dissertation as a book from the get-go. You told me that I needed to tell a compelling story, and that's what I've worked hardest to try to do.

Chris Smith, we've had many a productive conversation in your office. There, you pushed me to think more deeply about the cultural dynamics that color the landscape of opioids and pain. I hope I did our conversations justice.

Henry Jenkins, who took me on last minute as his advisee, also shared with me that, sometimes, it's better to ask for forgiveness rather than permission. Well, that's what we did, and it worked!

Sarah Banet-Weiser, my advisor and friend, thank you for letting me choose this as my dissertation topic. I know it was a drastic departure from what I had initially planned to write, but you supported me, nonetheless. And even though I still don't always believe that the rules apply to me, I won't forget that it was you who let me spread my scholarly wings. Chapter 3 is for you, sister.

And Andy Lakoff, who was (and continues to be) a brilliant advisor and one of the smartest people I know. Your intellectual contributions are spread out all over these pages. Thank you for so many an office hour discussing Foucault's biopolitics. I still have the notes I wrote then, and I refer to them often.

Outside of my committee, several other scholars provided me with advice and inspiration. I drew heavily on their experience and work when developing this project. They are Marcia Meldrum, Kelly Ray Knight, Eugene Raikhel, Helena Hansen, and Emily Martin.

Aside from researching, teaching, and writing a dissertation, I was lucky enough to form many friendships in the doctoral program at the University of Southern California. From each of these friends, I collected advice and drew inspiration. Thanks for that, Andrea Wenzel, Shabnam Shalizi, Dan O'Reilly-Rowe, Anna Loup, Nathalie Maréchal, Lik Sam Chan, Katie Elder, MC Forelle, Perry Johnson, Courtney Cox, and Kristen Steves.

After graduate school (and a stint in Brazil), I moved to Manhattan to work at New York University with Eric Klinenberg, who, during my postdoctoral appointment, graciously gave me both the time and encouragement I needed to revise the manuscript. The method of social autopsy is one of several of your ideas that animates these pages, and which I believe improved the book.

I'd be remiss not to also thank my colleagues at Knology in New York City, especially Rebecca, Jena, and John. Your friendship and solidarity give me excellent reasons to get to work, and you've cheered me on, on the many occasions when I've referenced this project.

Then, I must also thank the anonymous reviewers who read and reread this text, and who made insights and suggestions that greatly improved it. And a big thank you to my editors at Oxford University Press in the United Kingdom and New York, who brought this idea to fruition.

Outside of work, Mom and Dad, my sister Marisa, and my grandparents were crucial in providing me with limitless love and support. Without them, all of this would have been but a dream.

Somewhere between Barcelona and California is my dear friend Manuel Castells, who always believed in this project and urged me to keep going, even when I had decided I would never write or think again. *Te lo agradezco, mucho.*

Here in Brazil, where I currently live, I am motivated by my two fierce stepchildren, Luiza and Rafael. Your unbridled enthusiasm for politics, history, and human rights inspires me more than you will ever know.

And finally, I owe this book to my husband, Marcelo. How many hours have you spent listening to my digressions on the politics of pain, the many meanings of health, and my hopes for the future of drug policy? You pushed me even when I was unwilling to be pushed and have been the best sounding board I could have ever asked for. *Obrigada, meu companheiro, meu lar, e meu amor.*

Opening

In early 2018, *TIME* magazine published an online photojournalism series on what has become known as the U.S. "opioid epidemic." "The Opioid Diaries" requires visitors to scroll down through a series of black-and-white photographs, each of which is accompanied by captions and quotes. In the first image, situated at the very top of the page, a man and woman are photographed crouching in the back of a van, which the caption says is parked in San Francisco. Inside it, the couple balances on their shins as they aim needles full of clear liquid at the veins they've found in each other's arms. Further down the page is a pick-up truck with a man named Chad reclining in the driver's seat. His body is lax and, although his eyes are open, his head is cocked backward so far that it nearly escapes the frame. He overdosed shortly before the picture was taken and now lies in a state of semiconsciousness as a trio of paramedics leans through the truck's open door and try to revive him. Below Chad's image is a photograph of two people—a elderly woman and a young man—embracing inside a funeral parlor in New Hampshire. The woman's 24-year-old granddaughter, Michaela, can be seen in the background lying colorless and quiet inside her open casket. Scrolling down, a quote appears in large block letters on the screen; they belong to an Ohio sheriff who, commenting on what he's seen of the opioid epidemic in his own state, says simply, "It's nonstop. It's every day."

Continuing down the page, a woman in Boston squats on a mound of dirty snow. Whether because of overuse or the dropping temperature, she is struggling to find a vein. Below her, a homeless grandmother injects heroin next to a shopping cart full of her belongings. She is followed by a series of police officers. Each is accompanied by overdose victims, whose limp bodies were found on the side of the road in New Mexico, or stretched out on a bed in Ohio, or crumpled in a heap on the kitchen floor. With every movement of the mouse, more and more photos appear on the screen, more and more people, though most of them are bodies—depictions of death found in every corner of the nation, from Massachusetts and New Hampshire to Florida, Ohio, Kentucky, and New Mexico. But finally, at the bottom of the page is an image of life—a tiny baby, born just seconds before the photo was a taken. On first glance, the photograph is a hopeful one, though the quote accompanying it tells a different story. For this little girl is but one of the many children who, from the moment they enter the world, are destined "to be born addicted." She, and the others like her, are not symbols of hope, nor of life, but of a bleak and provisional endpoint: The first book in "The Opioid Diaries" may be full, but another one sits on the shelf beside it—a stack of blank pages waiting to be filled by the next generation of corpses-in-waiting and the stories they'll tell about a future where nothing is getting any better.

The story offered to us in the "The Opioid Diaries" not only describes a desperate reality located in the present but also lays out a nightmarish trajectory of the future—a story set in motion by the descent of so many people into the labyrinth of addiction. This dark space that, as all of the images described above suggest, is occupied by ever-expanding networks of drugs, death, and, perhaps most tellingly, of *pain*. The networks of pain that encompass people who use opioids, their families, as well as social workers, counselors, and police officers, among others, have been growing exponentially over the past 20 years and yet show no signs of stagnating—at least not anytime soon.

But while "The Opioid Diaries" and many other striking news stories like it point to opioid use and addiction as the key problems that define the current crisis, I disagree. I want to suggest, as this book does, that understanding opioid epidemic—what it is, how it evolved, and how it gained meaning and significance among the American population—is a story that is better told through the lens of pain. After all, it was pain and, specifically, the various moves that rendered certain pain medications legible as safe and popular drugs of choice that helped us get here in the first place. Borrowing a term used in popular parlance, it often seems as if we're being pushed down to "rock bottom," while growing pain networks continue to encompass more and more of us and the people we love.

It is no stretch to say, as many already have, that the ongoing opioid epidemic represents the worst overdose crisis in modern history—one that, over the past two decades, crept slowly but surely into the lives of millions of people living in North America. The rate at which Americans and Canadians are now overdosing and dying from the use of opioid drugs is astonishing: Since 1999, U.S. overdose deaths involving opioids have quintupled, to the extent that drug overdoses now constitute the leading cause of accidental death in the United States (CDC, 2017a, 2017b, 2017c). According to the most recent CDC estimates, in 2021 alone, overdoses killed nearly 108,000 Americans—a tally that far surpasses the total number of fatalities caused by car accidents and gun violence combined (Ahmad et a., 2022). Opioids, which account for 75% of all drug overdoses, are the principal force driving up this body count.

While opioid overdoses were climbing throughout the 2000s, so too was the problem of pain. The number of pain diagnoses from 2000 to 2009 show a gradual, yet stable, increase of about 3.3% and then suddenly shot upward, marking a threefold increase from 2009 to 2010, with an even greater wave seen in chronic pain diagnoses, which escalated 4.4-fold during those same years (Murphy et al., 2017). Today, according to the CDC, more than 50 million Americans suffer from chronic pain. That's about one-fifth of the adult population and includes 20 million people who suffer from what's referred to as "high-impact" chronic pain, that is, pain that limits their life in such a way that they can't perform even the most basic daily functions (Dahlhamer et al., 2018).

But looking at these parallel trends in opioids and pain, key questions remain: What kinds of social forces underpin these numbers? How did it happen that so many people in pain began seeking out, receiving, and consuming opioids? After all, opioids represent just one of many classes of pain-relieving medications, including many over-the-counter options, nonsteroidal anti-inflammatory drugs (NSAIDs), nonopioid analgesics, antiepileptic drugs (AEDs) tricyclic antidepressants (TCAs), local anesthetics (LAs), and more. How did it come to pass that opioids—the class of pain relieving drugs with the longest and perhaps most well-known history of dangerous side effects, overdose, and death—suddenly took their place in the medicine cabinets of so many people in pain? This problem of pain, which cannot be separated from the waves of opioid supply and demand, is precisely what this book seeks to understand.

The Problem of Pain

As I dug deeper into the opioid epidemic—its images, narratives, and the stories of people who take opioids as well as those who manufacture, prescribe, regulate, and police them—I realized that I couldn't answer any of the questions I had about opioids without running into some kind of question or issue concerning pain. And in the years I spent trying to learn everything I could about these drugs—which meant sifting through decades of medical journal articles, opioid advertisements, U.S. Food and Drug Administration

(FDA) meeting transcripts, court cases, congressional hearings, and drug forum threads, among other things—I kept running into institutional discussions and debates around opioids that either revolved around, questioned, or were somehow stalled in the process of trying to understand how pain works and how best to deal with it.

In medicine, for example, arguments regarding the possibilities and pitfalls of using opioids as a pain treatment were waged among those with differing approaches to the problem of how to adequately measure and treat the problem of patients' pain. Those who saw pain as a "vital sign," an objectively measurable object equal in importance to one's heart rate, breathing rate, and temperature, tended to advocate on behalf of opioid-based pain interventions, while those who approached pain as a behavioral symptom of a more complex social problem were often opposed to the uptake of opioids in treating pain.

Outside of the medical domain, the problem of pain has also been a powerful tool in the cultivation of consumer demand for opioid products. For instance, consider the rapid uptake of the now notorious opioid OxyContin, which was made famous as the painkiller that jump-started the current crisis. Many have attributed this uptake to the wide-reaching and incredibly aggressive marketing campaign that its manufacturer, Purdue Pharma, launched in order to sell it. And while I certainly do not disagree, what I found in my examination of Purdue's campaign is that the successful branding of OxyContin was better explained not in terms of the promotional claims (many of which have since been outed as lies) that were made about the drug itself, but in terms of the way in which Purdue branded the problem of pain by defining it as the antithesis of independence, self-reliance, and the American dream.

In every dimension attended to in this book—which includes the evolution of opioid use in medical practice as well as in pharmaceutical marketing, opioid regulation, the history of opioid addiction, and popular culture—pain is the linchpin. In particular, the way pain has been constructed within and outside medicine is key to understanding how and why opioids became so widely prescribed in the United States. The social construction of pain is what has led us to the so-called "opioid revolution"—a phrase that has been used to discuss the rapid popularization of opioid-based treatments in medicine and their normalization in culture, more generally. On the political stage, pain has informed calls made by politicians, patients, and advocates alike to break away from the United States' historically punitive stance toward opioid use and opt instead for a more compassionate approach to the problem.

In a number of recent political speeches, pain is the affective link through which politicians accrue support and promote platforms that commit to putting an end to opioid overdose, to the misery that accompanies it, and to the escalation of drug-induced fatalities now being rendered as "deaths of despair" (Case & Deaton, 2015, 2017). Through the pathos of pain, the opioid epidemic has itself becomes legible not so much as a criminal scourge but as a national tragedy, which is continuing to evolve within American culture today.

Finally, pain is the lens that forces us to interrogate the logics, protocols, strategies, and structures that make our health system function yet that have also contributed to the current crisis. In opioid regulation, for example, drug regulators' decision to fast-track opioids and curb rising rates of pain diagnoses was made in light of regulators' self-expressed doubts about how to define and measure pain. This uncertainty, when combined with the United States' dominant regulatory logic of ignoring potential risks until the moment in which they can be quantified, laid an extremely fertile ground out of which the current crisis could grow.

But to say that the opioid epidemic is more about pain than about opioid drugs is not to sideline them completely. It is instead a way of coming to see how pain and opioids became so tightly connected. Opioids, to be clear, represent a class of drugs that includes substances derived from the opium poppy (including morphine and heroin) as well as synthetic and semisynthetic drugs that act on opioid receptors in the brain.[1] These include both prescription drugs such as codeine, oxycodone (OxyContin), and hydrocodone (Vicodin), as well as illicit substances such as fentanyl and its more powerful analogs, like carfentanil. Every year, millions of Americans use these drugs: In 2020, 9.5 million Americans reported misusing opioids (SAMHSA, 2020). And despite various efforts on the part of regulators and law enforcement to crack down on opioid use at local, regional, and federal levels, these numbers continue to climb. The extent to which opioids have penetrated American lives is, frankly, incredible. According to a 2016 survey released by the Kaiser Family

---

[1] It may be worth noting that, for technical purposes, opioids are divided into two categories: Opiates represent all drugs that are naturally produced from the opium poppy, which include substances such as opium, morphine, and codeine. Opioids, on the other hand, is a broader term (which includes opiates) that also refers to substances that are either synthetically or partially synthetically produced (such as oxycodone, fentanyl, hydrocodone, methadone, heroin, and others) and bind to the brain's opioid receptors. For the purpose of clarity and simplicity, and because this book concerns both types of substances, the word "opioid" will be the term that is primarily used throughout these pages.

Foundation, roughly half of all Americans said that they personally know a friend or family member with either a current or past addiction to prescription opioids (NCADD, 2016). And there are many others, no doubt, who know someone using opioids outside this sanctioned medical domain.

It is striking that opioids have come to wreak such disastrous havoc in the lives of so many people in such a short amount of time—or that in 2014, in 12 different states, the number of opioid prescriptions far outnumbered the number of people living in that state. These numbers tell a story about a nation in pain, where one in five adults are diagnosed with a chronic pain condition, and where constant suffering, for them and many others, is the norm.

Equally startling is the fact that these drugs, which we have known for thousands of years are potent, came to occupy space on the shelves in our medicine cabinets. That is to say, it is telling how their meanings have shifted, how they have, over time, been redefined as cures rather than poisons and as first-line medications rather than dangerous "drugs."

While the opioid epidemic is not the first overdose crisis to plague the United States, it is novel in a few important ways. The first has to do with its scale: Between 1999 and 2019, nearly half a million people died from an overdose involving an opioid (CDC, 2021)—a number that outpaces the number of American soldiers killed in battle during World War II (DVA, 2017). And it is no small thing that by the time I finish writing this introduction at the end of this particular Friday, at least 130 more people in this country will be dead, and their autopsy reports will say that it was opioids that killed them (NIDA, 2020).

Second, the opioid epidemic is unique in that it represents an overdose scourge that has involved not only illicit substances, but also legal ones—chemical compounds produced by pharmaceutical companies, regulated by the federal government, prescribed by doctors, and distributed through official channels within the U.S. healthcare industry. In many ways, as this book will argue, it is precisely the dynamics of the legal spaces in which opioids have circulated that contributed to their rapid uptake and to the eventual opioid surpluses that saturated the pharmaceutical market and caused these drugs to spill out of their legal context into illicit and gray drug markets. In these unregulated or semiregulated spaces, synthetic opioids are now produced on a massive scale in laboratories, advertised through Internet platforms, and distributed through the U.S. mail system and at brick-and-motor pain clinics that now populate many of the strip malls that line this country's streets.

The fact that shadowy gray and illicit markets are intertwined with the legal market for prescription opioids points to another key characteristic of the current epidemic, which is the threat that its existence poses to the legitimacy of U.S. health institutions and to law enforcement agencies and regulatory administrations, all of which are now charged with intervening in a problem space that they helped to create. Put another way, the opioid epidemic sheds light on a fundamental paradox: The institutions that we would typically hold responsible for identifying solutions to this problem find themselves unable to do so. And it's no surprise, considering that the very logics, norms, and procedures that bolster their legitimacy and define their functions have also enabled the unchecked expansion of pain networks.

An additional characteristic of the opioid epidemic is that it was framed as a problem that was overwhelmingly white. The "whiteness" of the opioid epidemic has conditioned many of the tangible ways in which the health establishment and criminal legal system have chosen to intervene in drug use and on people who use drugs (PWUD), just as it has also been a crucial component that has shaped the public discourse surrounding them. To provide one example, the construction of opioid use as a criminal issue requiring punishment and imprisonment—which was a dominant approach to the use of heroin among people of color throughout the 1980s—now exists in tension alongside other discourses, some of which advocate for a public health and human rights approach to drug use. Today, we are likely to find speeches and political positionings related to opioid use that take a "compassionate" stance toward it, opting less for incarceration than for rehabilitation. This narrative is paralleled by the construction of opioid use as an overwhelmingly white problem and as something that warrants empathy rather than punishment. Indeed, the subtexts of race and class—specifically of whiteness—are embedded in many of the discussions and protocols that have been used to intervene in the problems of pain and opioid use. Much has been made of the fact that for the first time in a generation, the life expectancy of white Americans is declining, presumably due to painkiller-related overdoses and suicides. Moreover, race and class shape the political and institutional positions that condition the ways in which drug use is interpreted and experienced by different segments of the American population.

Pain and Uncertainty

On October 26, 2017, then President Trump signed a presidential memorandum ordering the Secretary of Health and Human Services to declare the opioid epidemic a "nationwide public health emergency" (Hirschfeld Davis, 2017; Wagner et al., 2017). But what makes a situation an emergency, or an "epidemic"? Uncertainty and unpreparedness are problems that are central to the construction of the opioid epidemic as both an epidemic and a public health emergency. We were, in many ways, unprepared for the opioid crisis. We were, for reasons that will be laid out in Chapter 2, poorly equipped to anticipate the extent of opioid use that would come in the wake of approving new opioid products. And we have since been uncertain as to how to deal with the consequences of these decisions.

And after all, what is a declaration of a "public health emergency" if not a very public manifestation of panic in the face of extreme uncertainty? In particular, the uncertainty regarding the relationship between opioids and pain is one that has long plagued medical professionals, regulators, patients, and politicians. It has dominated their various meetings and discussions, ultimately anticipating a host of hasty decisions, undergirded by weak or nonexistent knowledge (which these experts themselves admit to) about how to measure the intensity of pain, to what extent opioids can be used to treat it, how to measure their effectiveness, and how to know whether or not they pose a risk to pain patients.

The discourse of uncertainty in the face of pain management, opioid development, regulation, marketing, and treatment is absolutely crucial to accurately telling the story of the opioid epidemic. It is important in part because it is a discourse that anticipates the use of particular kinds of interventions that are designed to manage it (Lakoff, 2017). In the opioid crisis, interventions that have been utilized in the face of uncertainty about a present predicament have tended to rely on historical data. And they have used that data to attempt to anticipate the evolution of the crisis. Regulators and scientific experts attempted to overcome their uncertainties regarding pain and the risks of opioids through the mobilization of two major techniques: (1) the production and promotion of new "life-saving anti-abuse" technologies and (2) surveillance, or "pharmacovigilance" techniques, explicitly designed to identify and mitigate health risks as they unfold in real time.[2] Both types of intervention, which were developed and widely lauded in the name of eradicating uncertainty and ending opioid abuse, were developed relying on historical data about similar, but not identical, patterns of drug use. In doing so, they made the mistake of neglecting to incorporate the novel realities of

---

[2] As I'll discuss in detail in Chapter 2, the use of pharmacovigilance and similar surveillance techniques in opioid regulation can be understood as a response to a horrible system failure—in

opioid use and, as a result, had incredibly negative effects both on opioid consumption trends and on the ability of regulators and law enforcement to anticipate and manage new kinds of risks associated with opioid use.

As a public health emergency rooted in extreme uncertainty and a series of failed attempts to remedy the problem, the opioid epidemic is also a reflection of widespread public concerns and doubts regarding the promise of scientific expertise, pharmaceutical innovation, and the usefulness of our nation's health system. As are all public health emergencies, the opioid epidemic is emblematic of many of the shortcomings of institutional expertise in dealing with seemingly unmanageable risks. It is, in Lakoff and Collier's words, a product "of the scientific frameworks and governmental practices that seek to know and manage" it (Lakoff & Collier, 2008, p. 7).

Perhaps what the public has found, in expressing such anger and disappointment at the public health establishment and other associated actors, is their creation of and tepid response to something that Ulrich Beck (1992) referred to as "modernization risks"—risks that not only give rise to emergencies but are actually produced from within the workings of modernization itself, often by the very same institutions, knowledges, and practices that were meant to contain them. Put another way, the opioid epidemic is a problem that did not emerge out of thin air. It should instead be understood as something that grew out of the specific knowledges, practices, and logics being applied from various domains of expertise within our health-related institutions (medicine, pharmacology, regulation, accreditation, etc.)—those same institutions that Americans have relied on for the promotion and protection of their health, comfort, and safety. In recent years, these experts and institutions have emerged as key contributors to the current crisis and, in many media accounts of the epidemic, are disparaged for their irresponsibility in failing to predict it as well as for their subsequent failure to respond to it (Frydl, 2017; Von Drehle, 2018).

One example of expert knowledge that appears to have collapsed under its own weight in the opioid crisis is the logic of medicine taken "as prescribed." This logic, which is most often expressed as a directive in medication guides, suggests that a given medicine will work as it should so long

---

which the FDA attempted to regulate a particular opioid medication using an older approach to public health (one that relied on historical data and analogies of patterns of use with "similar" kinds of drugs), which would end up producing a serious gap in knowledge, and which prevented the FDA from predicting the soon-to-be surge of opioid overdoses that, in some accounts, represent the start of the entire epidemic.

as it is taken in the proper way—so long as patients use it "as prescribed." By extension, when a patient takes their medicine in a way other than "as prescribed," they are seen as having put themselves at risk—and the risks are unknown. The rule of "as prescribed" has a number of key functions within the U.S. health system: First, it serves as a safety measure, which helps fulfill the FDA's basic regulatory tenets—safety and efficacy. Second, it provides a legal logic that protects regulators, manufacturers, physicians, and pharmacists from prosecution should a patient suffer an injury or die as a result of taking their medication not "as prescribed." Third, it provides a means of classifying individual consumers as well as their consumption behaviors: Those who take their drugs "as prescribed" are patients, and their behavior is neutrally understood as medication "use." Those who deviate from the rule of "as prescribed" take on a different status; they are not patients but addicts, whose consumption practices are reconstrued as "abuse."

The usefulness of the logic of "as prescribed," which has long attempted to control and police pharmaceutical drug consumption, is breaking down. We see this in media accounts of the opioid crisis that relay the stories of patients who took opioids just as their doctors ordered yet still overdosed and died—their loved ones saddened, confused, and uncertain about how and why this could have happened under a physician's care. We've seen such stories written in bold in headlines time and time again. And we've seen it in the panic that corresponds to these stories and in the anxieties of professionals and the public alike, which worry that far too many patients are transforming into addicts (Calabressi, 2015). So too do we read stories that signal a sharp escalation of fear and doubt—doubt that is expressed with regard to medications that have killed patients, just like a "drug," or a poison, would do. And there are other narratives, some of which tell us that doctors are no longer doctors but drug dealers; these narratives are supplemented with news about cases launched against doctors (including the physician of the late musician Prince) whose patients died from taking their medications "as prescribed" (Deprez & Barrett, 2017; Moshtaghian & Meilhan, 2018; Pazanowski, 2018). In the opioid epidemic, we can see how "as prescribed" and "not as prescribed," "use" and "abuse, medications and "drugs," and other similar dichotomies are breaking down, both in media accounts of the epidemic and in the official statements made by government officials and health experts who, while not exactly admitting anything, hint that the problem is, by now, too far out of their reach.

The "Grayness" of Pain

The dissolution of familiar dichotomies or boundaries in both institutions and culture shapes the subjectivity, or self-understanding, of people who use opioids. As discussed in Chapter 5, the distinctions that separate "patients" from "addicts" and "at risk" people who use opioids are made and repeated (in both the mass media and in institutional discussions) precisely because of the systemic uncertainty that undergirds the opioid epidemic and the problem of pain. Over the past few decades, crucial questions and doubts about the nature and measurement of pain have given rise to questions about pain treatments, which in turn have given rise to widespread confusion and contradictions regarding the types of people who use them.

But why do pain and pain patients seem to augment the importance of boundaries and categorizations? My provisional answer to this question is that pain is a particularly gray area. Part of pain's grayness has to do with the fact that it is a highly subjective and heterogenous experience that has come to be understood through objective and homogenous measurement techniques—a contradiction that has made it difficult to resolve many of the questions that surround it. How do we really know whether someone is in pain? When does pain become severe enough to warrant opioid-based treatment? How can we know that pain is being adequately managed? How many people experience pain, and how do they experience it? What systems of measurement and categorization do we have in our scientific, regulatory, and legal toolkits to pose answers to these questions? In the opioid epidemic, this "grayness" is everywhere and underlies both the ways in which pain is talked about and the kinds of interventions that have been developed to deal with it.

But the grayness of pain is not new. On the contrary: These slippages and categorical breakdowns within the opioid crisis have long been central to the problem of pain. They have haunted the field of pain management for decades and have prompted it to continuously grappy with uncertainties regarding pain's definition, measurement, and treatment. Throughout the past half-century, chronic pain experts have been engaged in the task of finding appropriate methods to objectively assess what is now broadly known to be an inherently subjective experience. Doing so, after all, has been understood as necessary both for assessing the effectiveness of opioid treatments and figuring out where to draw the line in opioid prescribing. Like scientists and medical professionals, regulators have likewise been confronted with these same questions, which they need to answer in order to assess the safety and efficacy of opioid products as well as to decide whether to approve new pain medications. The objective assessment of subjective pain has been a crucial stumbling block for the FDA, which has often found itself paralyzed by pain's complex reality. Moreover, the grayness that challenges scientific and regulatory experts to provide confident answers to the question of pain has amplified as opioid overdoses have continued to rise. The skyrocketing rates of overdoses related to drugs that were originally deemed to be abuse deterrent or safe have only augmented this problem, making it all the more difficult for medical practitioners, regulators, and other healthcare professionals to properly assess the benefits and risks of opioid medications.

Grayness is likewise present when one examines the opioid markets. The opioid crisis evolved in different market contexts, including one that is legal, regulated, and institutionally channeled. This legal context is imagined to be distinct from the "black market," where the manufacture, distribution and consumption of opioids are more clearly legible as illicit. But when you have a substance like a prescription painkiller that is legally manufactured and regulated by the FDA, but that circulates as much through informal exchanges among friends and family members as it does through a physician's written prescription and can be consumed in ways that are not "as prescribed," how do you classify it? Is it still a medication, or is it a "drug"? How can we characterize the everyday, informal contexts in which these slippages take place?

The contexts in which medications, drugs, patients, "addicts," people who use opioids, and other such categories are gradually breaking down constitute what I refer to as the "gray market" for painkillers. By "gray market," I am referring to a domain that is marked by practices that trouble the traditional mechanisms of legal opioid production, distribution, and consumption. In the gray market, these mechanisms also work to undercut many of the key institutional logics of drug regulation and enforcement, such as the logic of "as prescribed." The gray market is also amplified by those same logics that characterize opioid prescribing, marketing, and regulation. Because they do not always work the way we think they do, these logics begin to drive a rapid cycle in which uncertainties lead to dire consequences. For example, as Chapter 2 demonstrates, the doubts that regulators had about what pain is and how to measure it and its treatment played an outsized role in the decision-making of regulators and other key decision makers, which ended up pushing legal opioid prescriptions into illegal or, at the very least, more shadowy domains—that is, into gray markets where the drugs that circulate are legal ones but are distributed and consumed in ways that are not, which makes them particularly difficult to regulate and control.

The emergence and growth of the opioid gray market has relied primarily on two key processes, both of which are dedicated their own chapters in this book: branding and regulation. The branding of opioids has played a decisive role in constructing their meaning as a normal part of everyday life, which has helped to render their circulation in the gray market as something that is neither purely legal nor illegal, neither medical nor criminal, but somewhere in between. The processes through which the branding of opioids familiarized and domesticated them likewise helped to transform the meaning of these drugs, from something that was previously understood as a dangerous narcotic into a first-line medical treatment and staple in medicine cabinets across the country.

Part of the steady domestication of prescription painkillers before and throughout the opioid epidemic has much to do with the way in which these drugs have been connected, in advertising, to the problem of pain. Specifically, it has to do with the fact that branding opioids relied heavily on branding pain as antithetical to the American dream. That is, pain was incorporated into opioid branding not only as a symptom but also as an enemy of self-realization and the individual pursuit of happiness and other American values. Rewriting pain in this way has gone hand-in-hand with the uptake of opioids among doctors and patients. The extensiveness of this uptake, which led to sharp increases in opioid production quotas, has resulted in a pharmaceutical market overflooded with huge amounts of opioid drugs, many of which eventually find themselves diverted into the gray market.

This inundated opioid market and the presence of these drugs in so many American households are bound up with shifts in the contexts of their circulation. Once in overwhelming supply, many of these drugs moved out of the pharmacy and into less regulated domains. They were borrowed and gifted among family members, lent and exchanged among groups of friends. In this unregulated gray market, drugs are not "copped" on the street, nor "purchased" at a pharmacy; instead, they are exchanged in situations that are far more ambiguous in terms of their legal status.

The gray market for opioids also has implications for the people who participate in it, whose definition escapes the traditional patient/addict dichotomy. That is, if a young man with a muscle sprain borrows a few of his mother's pain pills and either takes them himself or hands them over to a friend who has a bad toothache and no prescription, Is the friend just a friend, or is he a dealer? What kind of opioid consumer do we consider his friend? A patient? An addict? Has one of these people committed a crime? Can they be punished? Should they be? These questions are not easily answered, neither by regulators nor by lawmakers, and they are difficult to interpret in these terms.

Ultimately, the uncertainties of the gray market, as well as the link it forms between the institutional and the illicit, is reshaping the broader landscape of drug use and contributing to a corresponding crisis of legitimacy within America's health institutions. We can see an example of how this has happened when we look at so-called "abuse-deterrent" opoioids, which regulators approved to replace older, more "addictive" pain medications. They did so, however, without understanding the effects such a change would have on the terrain of opioid use. What happened, and what the FDA either did not or could not foresee, was that opioid consumption would not be deterred so much as it would be reconfigured, by both alternative drugs and a variety of creative inventions used to deactivate the antiabuse mechanisms of their new-and-improved opioids.

In the wake of this antiabuse measure, patterns of opioid use largely shifted away from the newly formatted prescription opioids and toward more powerful (and far deadlier) synthetic opioids—namely, illicit fentanyl. Fentanyl—because of the unique characteristics of its chemical composition as well as because of the difficulties in regulating the channels through which it is manufactured and distributed—poses a grave threat to regulators and law enforcement, whose current tactics for policing illicit drug use are ill equipped to combat such a novel phenomenon. In this case then, the FDA's attempt to keep its products within the legal prescription drug market caused a huge and rapid slippage of opioid use toward "fentanylization" and the ever growing crisis of fentanyl overdoses that characterizes the opioid epidemic today.

Overview of the Book

This book seeks to characterize and analyze ongoing public discussions and debates about pain and opioids in the midst of the ongoing "opioid epidemic." To do so, it draws upon a qualitative and interpretive method of analyzing texts that is broadly referred to as discourse analysis—a form of textual analysis that takes into account the purposes and effects of different types of language, how beliefs and values are communicated, and how these beliefs are generated in ways that reflect or connect to the broader social and political context in which they are espoused. In particular, this book develops a discourse analysis of the communication landscape around opioid use and attempts to link what has recently been debated in pain management, opioid regulation, and pharmaceutical branding to the broader landscape of pain in U.S. society.

The method of discourse analysis applied throughout this book's chapters relies on the compilation of a range of cultural and institutional texts, which originate from traditional media outlets and social media campaigns as well as outlets dedicated to particular institutions, regulatory organizations, pharmaceutical companies, and advocacy group websites. These texts, together with public reports, government documents, and marketing material that discuss opioids, opioid use, and/or people in pain, can be thought of as constituting the public discourse that relates to these issues. For example, one area of public discourse on which this book focuses is the historical and ongoing struggle in the news and among physicians and regulators to distinguish opioid patients from opioid "addicts." In analyzing the significance of this discussion and the effects it has on people in pain, I pay attention both to the ways in which language is used to formulate these debates, and how these debates are being structured in ways that point to broader cultural tendencies and social norms around drug use, opioids, and pain.

The chapters in this book can be understood as addressing the opioid epidemic by conducting a social autopsy of it (Klinenberg, 2002). Such a method asks, "how did we get here?," and attempts to diagnose the key factors that have contributed to the crisis we now face. To do so, it attends specifically to the key historical processes, protocols, and institutional and marketing strategies that have transformed the problem of pain into a growing crisis of opioid overdose. The five chapters in this book show that as overdose numbers climbed, the framing of pain led to key decisions and practices that help explain how the opioid epidemic could have happened in the first place. While the first four chapters of this volume attempt to illustrate the routes through which we arrived at the present moment, the final chapter asks about meaning making in the opioid epidemic and tries to understand why the crisis looks the way it does today. It builds on the arguments advanced in the first four chapters to show how pain and the uncertainty that surrounds it have contoured the opioid epidemic and its subjects. These chapters elaborate on how uncertainty and the historical tendency to categorize good and bad people who use opioids have constructed specific roles for doctors, patients, and people who use drugs. In doing so, they also set up a framework for defining whose pain warrants attention and relief, and whose does not.

As a first step in conducting an autopsy of the opioid epidemic, this book begins by tracing the history of pain and the development of the field of pain management. Within these domains, it looks at the emergence of the "pain revolution," a sudden shift in prescribing practices in which opioids rapidly emerged as the hallmark treatment for managing different types of pain. Before the mid-1990s, attitudes toward opioids within the medical community were reluctant at best. Physicians were wary of prescribing what they understood to be extremely potent and addictive drugs and did so sparingly, usually as a last resort for patients suffering only the most agonizing forms of cancer-related pain. Yet, in the mid-1990s, the notion of chronic pain became an area of hot debate in pain medicine, and the idea took hold that many doctors were undertreating long-lasting pain, and that opioids were not nearly as addictive as previously thought (Meldrum, 2016). After these revelations, opioid prescriptions skyrocketed, quadrupling from 1999 to 2013.

The first chapter of this book focuses on key shifts in the medical knowledge and practices related to pain in order to understand how pain and opioids came to be linked together. In doing so, it also examines how forces outside medicine acted upon and in conjunction with medical expertise in ways that helped spark the opioid revolution. That is, policies enacted by politicians working together with pharmaceutical companies, physicians, and chronic pain advocates also worked to construct a fertile landscape in which the problem of pain would become a significant issue—not just for medicine but for society in general. Within this landscape and the movement to recognize the realities of chronic pain, pain management was reframed not only as a scientific field but also as a universal human right. The moral significance that pain management gained with its articulation in a rights-based discourse and movement was a turning point both for the field itself and for the future of opioids.

The rights-based discourse that became attached to pain provided a strong justification for a concerted push in both medicine and politics to develop new ways of measuring and treating it. The paradoxical task of objectively assessing an inherently subjective phenomenon was a serious conundrum for pain physicians. Their efforts to adequately measure pain, underlined by a movement of chronic pain sufferers, helped develop and put into general practice a new protocol that opted for an understanding of pain as the "fifth vital sign." As a vital sign, pain became prioritized as a part of common-sense clinical practice and as something that could be quickly and adequately measured in the short span of a clinical encounter. Thus, its assessment became routine practice across a variety of medical specialties, as did its treatment. Responding to a need to treat this now frequently diagnosed condition, opioids were quickly incorporated as an effective tool for remedying a complex, subjective, and social problem within an objective clinical encounter.

Though the genealogy of pain management points to crucial moments in the longer trajectory of the opioid epidemic, it is also the case that neither opioids nor any other pharmaceutical drug can enter the market and become a brand without first passing through the regulatory apparatus of the FDA. To that end, Chapter 2 investigates the regulatory logics that undergirded the FDA's decision to approve the release of new opioid drugs onto the market and its subsequent decisions to leave them there, despite growing awareness of the risks associated with their use. This chapter, which focuses on the regulation of OxyContin and abuse-deterrent opioid formulations like Opana ER, shows how the FDA's ability to anticipate and prevent the risks associated with them was undermined at nearly every step of its regulatory process, where extreme uncertainty about pain, along with strategic ignorance and an overreliance on industry alliances, rendered it blind to the lived realities of opioid use. Additionally, this chapter discusses the ways in which opioid regulation led to the formation of a gray market for opioids—a phenomenon that further debilitated the FDA's regulatory capacities and thrust it into an evolving landscape of synthetic opioid use that it could not hope to contain.

Chapter 3 expands the diagnosis of the opioid epidemic by tracing its movement into another space—the domain of pharmaceutical branding, where opioids have undergone a social and cultural process of normalization, one that transformed them from scientific/medical objects into sellable products. It looks at the branding of pain relief through a case study of the marketing of what is today the most famous of all prescription opioids—the painkiller OxyContin and its manufacturer, Purdue Pharma. Through an analysis of dozens of print advertisements, video campaigns, and other marketing techniques that Purdue used to promote its product, this chapter shows how a medical condition was reconstrued as a national problem, one that inherently hinders individuals' pursuit of the American dream of self-realization, autonomy, and productivity. By tying pain to the American imagination in this way, opioids became increasingly easy to incorporate into

American hospitals and households, where they were situated as familiar, domesticated objects.

This new framing of pain and opioids in American culture has likewise helped to bolster the gray market for painkillers as well as the environment of extreme uncertainty that underlies the question of pain. Within medicine and regulation, domains in which firm categories have long been employed to define and manage the world of drug use, certain boundaries and "truths" about this shadowy world have begun to erode. Narratives of the opioid epidemic are brimming with evidence of such categorical breakdowns, where patients become addicts, doctors become dealers, cures become poisons, and solutions give rise to new and greater problems. Each of these shifts in meaning reveals a crack in the larger classificatory system. And as a result, the need to reclassify and re-establish meaningful distinctions between different kinds of opioid use and people who use opioids has become all the more acute.

The fourth chapter in this volume examines the ways in which the discourses that mark the pharmaceutical branding of opioids have become integrated into and reworked in the nonpharmaceutical self-help industry, to which many people in pain often turn when they are attempting to stop using opioid medications. This chapter traces self-help's entrance into pain management through theories and programs of pain "self-management" and discusses how the rise of self-help in the domain of chronic pain has also paralleled the creation of a new kind of pain patient—the pain patient-expert. In both the pharmaceutical industry and the self-help industry, pain is constructed as the antithesis to leading a good life, which itself is cast in terms of individual productivity and the ability to restore and then enhance one's own capacities. Such a discourse, which relies on the individual person in pain as the sole "self-manager" of their symptoms, is fundamentally incompatible with the reality that pain is a deeply social phenomenon—with socially informed causes and, potentially, solutions. Pain cannot, this chapter argues, be framed as a problem for individuals, since individuals who experience chronic pain are, more often than not, deeply embedded within larger contexts of social isolation and social inequality, which augment their suffering.

Finally, Chapter 5 shows how the opioid epidemic has been made to look the way it does and how people who use opioids are constructed not just in medicine but also in popular culture. In particular, the fifth chapter focuses on attempts to restore order to what now appears to be a system of meaning in crisis. It does so by developing a critical discourse analysis of opioid use, which traces how different meaning-making practices have historically named, classified, and characterized people who use opioids as different kinds of persons—which I refer to as "patients at risk," "pathological consumers," and "consumed addict-victims." Each of the figures in this typology, as I show, along with the ways in which they are imagined, performs a specific function within the opioid epidemic, conditioning whose pain is seen as legitimate and setting the stage for a compassionate approach to opioid use that is selectively applied to only some people in pain.

Finally, the book concludes by considering the implications of its findings for the development of drug policies and future research related to medicine, pain, and their relationship to culture. In addition to several practical suggestions related to the regulatory process, I also insist that developing a rich understanding of the opioid epidemic can be achieved by considering the relationship of the crisis to pain, as well as to systemic issues within the United States's health care system.

The opioid epidemic is, without question, a multifaceted phenomenon, one that says as much or more about the transformations taking place within American culture and the relationships we have to the question of pain than it does about any particular medical truth. So, while there are many possible interventions that could be utilized within our health institutions, protocols, and practices (and which will be discussed in this book's conclusion), other ameliorative work could—and should—involve the broader meaning-making practices around pain itself. For pain is a problem that, like the opioid epidemic itself, is not about isolated individuals but is instead a hypernetworked, political and economic, social and cultural issue. In the pages that follow, I propose to analyze some of the key dimensions within the complex world of pain in the United States and the ways in which we have come to understand, define, and ameliorate it by taking a deeper dive into the opioid epidemic, pain's most urgent case study.

1
Tracing the Painkiller Revolution

In 2011, the U.S. Centers for Disease Control and Prevention (CDC) issued a press release that referred to prescription opioid abuse as a "silent epidemic." The numbers of Americans dying from prescription painkillers had been creeping upward for some time: According to a report by the National Institute on Drug Abuse (NIDA), overdose deaths quadrupled between 1999 and 2007, escalating from 3,000 to 12,000 each year. Compare this with the number of deaths from cocaine, which killed about 6,000 people in 2007, or heroin, which killed about 2,000 (NIDA, 2015). That prescription opioids, drugs that circulate largely through tightly regulated channels, had been killing so many people every year is striking.

When attempting to account for these surging numbers of opioid prescriptions, overdoses, and deaths, some of the most common explanations have looked to medicine—specifically, to an "opioid revolution," a widespread transformation in prescribing habits among practitioners who, the story goes, have become increasingly willing to prescribe opioid products for patients in pain. While evolutions in medical thinking surrounding the safety and efficacy of opioids have certainly played a significant role in conditioning their use, the principal force that has conditioned the evolution of the opioid crisis—as well as the "opioid revolution"—is not doctors but pain. Pain is the linchpin around which strategies have been formed, protocols designed, and decisions made that together transformed opioids into the gold standard treatment for a variety of medical conditions and conditioned huge surges in their supply and demand. Relatedly, what has also laid the foundations for the current crisis is the uncertainty that has and continues to surround the problem of pain. For the problem of pain has brought with it a series of key questions: What is pain? Is it a disease or a slippery slope toward addiction? Is it undertreated or overtreated? Are the people being treated for it patients or addicts? Does its treatment constitute a cure or a poison? When something goes wrong, are the doctors who prescribe these treatments caregivers or criminals? In the opioid epidemic, the answers that have been offered to these questions have narrowed the possibilities for pain management and, in clinical settings, have severely limited the ways in which it has come to be treated.

Coming to terms with the opioid crisis therefore requires first understanding how pain became understood as a medical problem and, in the process, how the problem of pain came to anticipate the use of opioids as the gold standard for treating it. To that end, this chapter presents a genealogy of the medical field of pain management, which seeks to identify the key uncertainties, questions, proposed solutions, and collaborations that emerged within and between medicine and politics and, in doing so, help us understand how opioids became the go-to treatment for pain, how this connection strengthened over time, and how it helped to set the stage for the unprecedented growth of an opioid market.

Genealogy is a kind of historical analysis that attempts to explain how we arrived at a particular present conjuncture. Borrowing from Foucault (1977a), it is a "history of the present," one that aims to uncover the hidden context and conflicts that color the past and reveal the ways in which they have come to shape the present. Genealogy provides us with a path to understanding the opioid epidemic as it exists in our current moment, one that is characterized by skyrocketing rates of opioid prescriptions and opioid-related deaths nationwide. Second, as an example of what Stuart Hall (1987) has referred to as "conjunctural analysis," genealogy does not assume linear evolutions in knowledge/practices but, rather, attends to the effects of a constellation of forces—the coalitions, collisions, and contradictions originating from a variety of domains (both within and outside medicine) that, in their interaction, have helped to shape our current moment.

In the present, pain is understood in medicine as a vital sign, equal in importance to heart rate, breathing rate, and other crucial measurements. What this means is that pain, as it stands today, is something that is routinely assessed, which opens the door to its routine treatment and, as we've seen, to the routine prescription of opioid medications. But this was not always the case. Not long ago, pain was not even seen as a condition in its own right. Instead, it was understood as a side effect, inseparable from the "real" wound or disease that gave rise to it. As such, pain was not something that earned its own separate assessment and management. The reframing of pain as a "vital sign" marks a crucial moment in the history of pain management, one that has had incredible consequences for opioid prescribing. Coming to terms with the opioid epidemic means analyzing this shift in medical thinking, and understanding both how and why pain's meaning changed, how it came to be understood in this way, and what kinds of consequences have followed in the wake of the reconceptualization of subjective pain as an objectively assessed "fifth vital sign."

However, a genealogy of pain management also requires broadening our historical field of vision beyond medicine, since pain is not and has never been just a medical object: It is also a social, cultural, political, and importantly, a moral problem. Particularly key to understanding the opioid crisis is seeing how pain came to be articulated within a moral framework and how a morally inflected vision of pain has helped to position it as a human rights issue (rather than just a medical issue). These developments have had drastic effects on the practice of pain management, particularly where opioids have been concerned. That is, pain-relieving solutions, including opioids, have come to be seen not only as medical treatments but also as moral obligations—ones that physicians have been compelled to employ as a means of securing the basic human rights of patients. This obligation has anticipated a host of legal questions for both patients and physicians. These questions and quandaries to which the problem of pain gives rise have been filtered into pain-related legislation and into new medical paradigms that have likewise served to guide decision-making in health-related policy and pain management in ways that conditioned outpourings in opioid prescribing and gave rise to the "opioid revolution."

The genealogy of pain management can be understood as unfolding through two overlapping historical shifts. The first of these shifts involves the moralization of pain or, put another way, the conceptualization of pain as a space for moral intervention, framed within a discourse of human rights. The gradual conceptualization of pain relief as a moral obligation and a responsibility belonging to both medicine and government opened the door for pain advocacy efforts targeting the undertreatment of chronic pain. The second shift alongside pain's moralization has entailed its objectivation—including the long and ongoing search for measures to objectively measure the subjective experience of pain (for both medical and political purposes). Pain's subjectivity called up a need for efforts to reposition pain within quantitative frameworks, as a number on a scale, which could be used to unilaterally diagnose, assess, and treat what is now known to be an inherently complex condition. The technocratic tools that came to be used to measure pain and assess the efficacy of pain treatments, such as the numerical pain-rating scale, purported to resolve the complex problem of subjective pain by rendering it as technological object—a vital sign that could be routinely assessed in a brief clinical appointment. In this context, pain was reframed as a technocratic problem, one that encouraged the use of technocratic solutions. Opioids became the foremost technocratic tool that would ultimately come to be used as the solution to resolve both the moral and medical problem of pain.

> I swear to fulfill, to the best of my ability and judgment, this covenant: I will remember that I do not treat a fever chart, a cancerous growth, but a sick human being, whose illness may affect the person's family and economic stability. My responsibility includes these related problems, if I am to care adequately for the sick . . . I will remember that I remain a member of society, with special obligations to all my fellow human beings. . . . May I always act so as to preserve the finest traditions of my calling and may I long experience the joy of healing those who seek my help. (From the Hippocratic oath, modern version, 1964)

That medicine has long been bound up with moral questions is nothing new. Even a cursory glance at the Hippocratic oath quoted above asserts that those who practice medicine must also commit to a kind of practice that extends beyond the lab or clinic, beyond the diagnosis and treatment of disease. The "special obligations" of the physician, as the oath suggests, positions them in two roles: They are, at once, a scientist whose value lies in the specialized knowledge, objectivity, and expertise that distinguishes them from their patients, and a caregiver who is valued in terms of their humanity and the oath that they have sworn to ameliorate the suffering of their fellow human beings. These discourses, which define the social role of the doctor as a scientist and a caregiver, are integrated in the Hippocratic oath and are particularly pertinent to the discourse and practice of pain management. In this field, the responsibility of attending to human suffering is more than a latent principle, a given, which all doctors know is expected of them. Here, suffering is the medical object itself. It is not a side effect of a specific diagnosis but diagnosis itself—one that has, over the past half-century, given rise to an array of theories, procedures, and interventions that together constitute the field of pain medicine and the focus of physicians who specialize in the management of acute and chronic human suffering.

Over the years, a historical and long-standing approach to the understanding pain in terms of its symbolic or spiritual value (as a symbol of life and strength of will or soul) has given way to a more pragmatic approach, in which pain is explained and examined not in relation to the soul or as the price to be paid for existence but in relation to the body and psyche. While coming to be understood in this way, as an embodied experience rather than an existential phenomenon, pain has also become an object of study—something to be researched, treated, prevented, manipulated, and controlled. It is this pragmatic approach to pain that underlies its transformation into an experimental object, and to the development of theories of pain in semantics, psychology, physiology, and therapeutics (Rey, 1995). The orchestration of these various disciplines, hypotheses, and theories paved the way for "modern" pain management.

In the 19th century, pain management swiftly evolved with the isolation of morphine and its industrial production in Germany in the 1820s. Another key development was researchers' widespread experimentation with different forms of anesthesia and the invention of the hypodermic needle, which simplified the administration of pain-reducing drugs. The needle also facilitated recreational self-administration, a practice that, as it became more widespread, aroused concerns about habituation and addiction (Campbell, 2007). In the United States, these concerns were addressed through the government's initial attempts at drug regulation and put into place in the 1914 Harrison Act, which regulated and taxed the production, import, and distribution of opiates. Not long after, the First World War normalized the medical use of narcotic pain medication. The use of new military technologies that resulted in mass casualties and grave injuries meant that medics relied heavily on morphine—not only to treat soldiers' injuries but also as an improvised form of anesthesia for emergency surgeries and amputations (Campbell, 2007; Tousignant, 2006).

Yet the history of pain medicine that matters most, for our purposes, begins during the Second World War with a man named Henry Beecher who, in the aftermath of battle, was charged with treating wounded soldiers who remained on the field. Beecher routinely offered morphine to the wounded who, to his surprise, often refused it. When he asked them about their pain, the soldiers reported shockingly low levels of discomfort. However, once they were removed from the battlefield and placed in the army hospital, their pain intensified, suddenly transforming from moderate discomfort to unbearable agony, even though their injuries had not worsened. What was this strange phenomenon, these patients' experience of horrific pain that, for some reason, seemed not to correspond with their injuries? The pattern Beecher identified did not fit within any of the previously established medical frameworks, which posited that "real pain" is always a proportionate response to a noxious stimulus (Baszanger, 1998). This understanding of pain as specifically tied to a wound or lesion could not explain what Beecher witnessed on the battlefield (Meldrum, 2007). He had no choice but to conclude from his interactions with wounded soldiers that the experience of pain was, in fact, more complicated than laboratory knowledge about it would lead one to believe. For, in fact, experiences of and reactions to pain are not necessarily proportional to the intensity of a given injury. Rather, the nature and mechanisms of pain are more complex. And pain's complexity, Beecher argued, would have to be studied not in terms of its relationship to a physical lesion, but in terms of the patient, the person who experiences and responds to it. This insight that pain is a fundamentally subjective phenomenon would inform the development of pain management for decades to come.

The assertion that pain is what the patient feels and thinks it is unfettered the scientific problem of pain from the laboratory settings in which it had formerly been confined. With experience now taking center stage in the study of pain, its theory and practice would have to move into new spaces, where experience could be analyzed and better understood. The study of pain management thus migrated into clinics. In the clinical context, pain research was to be conducted at the intersection of science and care, where it would necessarily center on individual patients and their self-reported experiences. A key assumption guiding the research was the need to take patient narratives at face value since "any attempt to evaluate pain must begin with the recognition that pain is a subjective phenomenon, and many factors influence the perception, response, and report of subjective events" (Jacox, 1979, p. 895). The consensus within pain medicine that emphasizes the importance of subjectivity would eventually (and somewhat paradoxically) give way to the quantification and objectification of pain in the clinical context. This process, as will be discussed in more detail, has been essential for both the formation of pain medicine as an established discipline and area of specialized expertise and for the narrowing of the practice of pain management, which is today centered on the use of pharmacological treatments.

In the United States, another physician was following Beecher's footsteps. John Bonica was an anesthesiologist whose interest in pain stemmed from his own experience dealing with chronic pain related to injuries he suffered during his former career as a wrestler. Like Beecher, John Bonica, who is now widely known as the leading pioneer of contemporary pain management, studied pain in the context of the war, treating soldiers who had recently returned home from overseas. Bonica was struck by the prevalence of soldiers who reported experiencing pain in the limbs they had previously lost in battle, even though the amputation sites were fully healed (Oral History Interview with John Bonica, 1993). This problem of lesion-less pain could not be understood using contemporary theories and often failed to respond to the treatments that were available at the time. Bonica was frustrated by the lack of treatment options, which were mostly limited to neurosurgeries (like lobotomies) and the administration of anesthesia. Both interventions were based on a mechanical medical model that saw pain as a symptom, which always signaled an underlying problem, most likely some kind of wound or lesion (Oral History Interview with John Bonica, 1993). Physicians were thus focusing less on pain itself than on the "real problem" that they were trying to identify and treat and to which pain was secondary. Yet in Bonica's experience, it was often the case that no "real problem" could be identified. It thus became clear to him that pain could no longer be understood as merely a symptom but needed to be seen as a diagnosis in its own right. What was also clear to him was that the diagnosis of lesion-less pain had a distinct set of psychological effects and that understanding how best to treat it would require the expertise not only of anesthesiologists and neurosurgeons, but also of psychologists and psychiatrists. Therefore, as Bonica asserted in his now famous book, *The Management of Pain* (1953),[1] the study of pain and its management would have to be a multidisciplinary effort.

In 1960, as his ideas about pain gained traction, Bonica was named head of the Department of Anesthesia at the University of Washington and provided with funds to set up a clinic in which he could put his theories into practice. Bonica's clinic, which he named the Multidisciplinary Pain Clinic, was structured on the basis of his assertion that pain is a multidimensional phenomenon and that advancing medical understanding of it requires specialists from different fields to pool their knowledge. At his clinic, practitioners gathered from a variety of medical disciplines, each with a different stake in the management of pain and armed with specialized methods for treating it. Especially key to the practice of pain management at the Washington Clinic was the influence of psychology—the science of subjectivity, a particular kind of expertise that was now at the center of pain's definition as "an unpleasant sensory and emotional experience associated with actual or potential tissue damage or described in terms of such a damage" (IASP, 1994).

---

[1] Bonica's book is now known as the first modern textbook of pain medicine and is regarded by many as the "bible" of pain management. See Baszanger (1998) for an extended discussion of Bonica's textbook and its reception within the medical community.

In particular, behavioral psychology was a critical component of many of the multidisciplinary pain clinics that had begun popping up around the country in the 1970s. The behavioral approach to pain management is generally attributed to the work of Wilbert Fordyce, a psychologist who was also stationed at the University of Washington, where he worked in the rehabilitation department at the university hospital. Fordyce, in an oral history interview he provided in 1993, explained that he had become frustrated with his lack of success in rehabilitating his patients, many of whom were heavily medicated and could not perform the physical tasks that their therapy required. Desperate for a solution, Fordyce decided to pursue the "hairebrained idea" of experimenting with a Skinnerian approach to rehab therapy (Oral History Interview with Wilbert E. Fordyce, 1993, p. 11). B.F. Skinner's famous theory of operant conditioning posits that the behaviors of an individual are a result of environmental stimuli, which, in an experiment, can be manipulated by the researcher in their attempt to alter an individual's behavior. With this in mind, Fordyce designed an impromptu experiment. He told his staff to ignore the complaints of their patients and to refuse them medication, which, he hypothesized, would only serve to reinforce and perpetuate their pain behaviors (Turner et al., 1982). Instead, the staff were told to give positive reinforcement only in response to their patients' attempts to exercise, socialize, and engage in other productive activities.

To his surprise, Fordyce's "hairebrained idea" worked. His patients' complaints became less frequent, and the patients themselves began to improve. In light of his success, Fordyce developed a behavioral approach to pain management that involved assigning to his patients a strict regimen of exercise combined with a schedule that gradually reduced the dose of their pain medication. When his patients stuck to the regimen, they received therapeutic attention. That is, the care they received was contingent on their good performance. This behavioral approach to pain management impressed John Bonica, who then invited Fordyce to join his Multidisciplinary Pain Clinic (Oral History Interview with John Bonica, 1993). As pain historian Marcia Meldrum has written, Fordyce's approach to pain management not only provided the field with a new treatment option but also contributed to a richer understanding of pain itself, for Fordyce and his colleagues had "completed the conceptual shift from *what real pain ought to be* to *what real pain actually is and does*." Real pain is defined by the patient. And the patient's experience of it would have to be a legitimate concern across medical disciplines (Meldrum 2007, p. 7).

In 1973, as interest in pain medicine was spreading throughout the United States and Europe, John Bonica and his colleagues decided to organize what became the first international symposium on pain.[2] The symposium was held in May that same year, in the remote area of Issaquah, Washington, where 350 people gathered to pool their knowledge about pain. Importantly, Issaquah brought together the folks working at Bonica's Multidisciplinary Pain Clinic with specialists from the British hospice movement and with a group of New York–based researchers who were conducting research treating cancer patients with opioid medications. Also present were Ronald Melzack and Patrick Wall, who in 1965 developed the gate control theory of pain (GCT), which is to this day perhaps the most widely known theory about the mechanisms of pain. GCT, which will be discussed in further detail later in the chapter, attributed the experience of pain to a spinal cord mechanism[3] and acknowledged the physiological processes that contribute to pain. In doing so, GCT helped to legitimize pain management as a field within medicine (Melzack & Wall, 1965; Wall, 1978).

By the mid-1970s, the field of pain medicine had already crystallized. A new medical journal, *Pain*, was established in 1975 and immediately began to publish the latest research on pain from around the world. That same year, IASP held its first world conference, which highlighted the importance of pain as a global public health problem and initiated the development of the first pain taxonomy, which was created to provide a "universal vocabulary" to healthcare professionals around the world.

The goal of universalizing pain management was also furthered by the development and proliferation of instruments and tools for "objectively" measuring the intensity of pain and the effectiveness of pain medications. For many years in pain management, a long and ongoing search was undertaken in search of a measure that would accurately account for the subjectivity and perception of pain. Among the measures that were developed and utilized within the field include the numerical rating scale (NRS), in which pain is rated on a scale from 0 to 10, the Visual Analogue Scale (VAS), in which pain

---

[2] The symposium jump-started the process of institutionalizing and globalizing Bonica's ideas about pain and its medical management. It is widely noted that this symposium laid the groundwork for the formation of IASP in 1974 and the development of the renowned medical journal *Pain* in 1975. A year after *Pain* published its first issue, IASP had already acquired 1,575 members from 55 countries who together represented 81 different fields of research (Baszanger, 1998).

[3] GCT posited that pain can be understood in terms of a spinal cord mechanism that normally blocks the transmission of painful stimuli to the brain by overwhelming them with a larger number of normal, nonpainful stimuli. Only stimuli of sufficient intensity can break through the "gate" and reach the brain, which then arouses the perception of pain.

is measured according to a 10-cm scale that extends from "no pain" to "worst pain," and the McGill Pain Questionnaire, in which pain is rated according to several, multidimensional questions, which assess the type of pain experienced (burning, pulsing, throbbing pain, etc.), how pain changes over time, the factors that increase the level of one's pain, and the strength of pain. The first two scales, which measure pain on numerical scales, gradually became the most frequently utilized in pain clinics and hospitals, in part because they lent themselves well to rapid assessment that would fit within the brief time span of a clinical appointment. As the use of these tools spread, they also became the foci of various sets of "guidelines" developed by the World Health Organization (WHO) and the American Medical Association, along with numerous other professional organizations. Situated within the broader field of public health, pain management also became part of routine medical practice outside of multidisciplinary pain clinics and pain specialists.

By 1979, a survey of pain clinics counted 426 around the world, with 278 (65%) of those located in the United States (Meldrum, 2007). Pain programs quickly spread throughout the country and were soon referred to as "medicine's new growth industry" (Leff, 1976). And so it was that some 30 years after John Bonica began attempting to "sell" the idea of pain medicine and a multidisciplinary approach to its treatment, his efforts finally bore fruit. And the modern discipline of pain medicine was born (Oral History Interview with John Bonica, 1993).

The Medicalization of Chronic Pain

In 2016, one in five U.S. adults reported suffering from chronic pain (Dahlhamer et al., 2018). Many of these individuals, moreover, have been prescribed opioids for their condition. Understanding how chronic pain emerged as a distinct public health issue is therefore crucial for coming to terms with the ways in which pain relief and, alongside it, opioids, became normalized in American society.

Numerous scholars have written about the medicalization of different kinds of social problems (Conrad, 2007; Conrad & Schneider, 1980/1992; Zola, 1972). Medicalization, understood as the process through which human problems become defined and understood as medical conditions, is central to the story of pain management and opioids. For it was not always the case that acute pain and chronic pain were understood as distinct medical diagnoses. But with the passage of time and the advancement of crucial theories about pain and its mechanisms, chronic pain eventually emerged as a proper diagnosis, one that expanded medical understandings of pain beyond seeing it as a byproduct of an injury or underlying illness and, instead, as something that must be measured and treated independently of these factors, on the basis of the patient's experience of it.

Different from acute pain, chronic pain can be understood as pain sensations that outlast a particular injury, extending at least three months beyond normal healing time (Treede et al., 2015). Chronic pain also includes pain that occurs in intervals, which comes and goes for months or years (Baszanger, 1998). While acute pain typically accompanies a specific injury or disease and often resolves over time, chronic pain does not. As Conrad and Muñoz (2010) have written, the recognition of chronic pain as a discrete medical condition shifted the focus of pain treatment from the elimination of pain to its ongoing management. This crucial shift, whereby pain was reframed as an ongoing, intractable condition, also anticipated the long-term use of opioids among chronic pain patients.

But how did this happen? A key pain theory, first published in 1965, is useful for explaining exactly how chronic pain stepped into the limelight as a public health concern. In their groundbreaking article published in the journal *Science*, Melzack and Wall (1965) critiqued the field of pain medicine for not contending with different types of pain. For until that moment, pain management had been primarily concerned with lesion-related, acute pain. It did not, the authors contended, spend enough time discussing chronic pain syndromes such as phantom leg syndrome and causalgia (chronic regional pain). Contrary to the rest of the field, Melzack and Wall were precisely concerned with these types of lesion-less pain. They argued, moreover, that different types of pain existed that did not have a physical cause. The argument the authors made on behalf of chronic pain crystallized in their now-famous GCT of pain, which suggested that pain sensations were not necessarily reactions to painful stimuli but, rather, were a result of pain signals in the brain. These pain signals, moreover, were modulated by different spinal cord systems that became activated by different stimuli. As a modulated phenomenon, pain often has no direct relationship to a particular cause. Instead, it has a variable relationship to pain sensations, which for Melzack and Wall are the key problems on which pain treatment must focus. Thus, with no underlying physiological cause, pain treatment would need to refocus its attention from treating pain itself to treating patients' reactions to and perceptions of pain. This crucial insight would inevitably shape the future of pain medicine, which would come to rely more and more on patients' perceptions of pain as measured through self-reports and self-assessments rather than on standard diagnostic technologies, like X-rays, that attempt to trace pain back to a particular physical cause.

With patient self-reports brought to the forefront of pain medicine as the key benchmark on which to base pain treatments, patients themselves took center stage in the practice of pain management, where pain would be reframed and assessed in terms of how the patient experienced, perceived, and spoke about it. Patient centeredness in pain management was also paralleled by the development of pain advocacy among physicians and patients, who attempted to increased cultural awareness about the problem of chronic pain and its undertreatment.

The emergence and acceptance of chronic pain as a distinct diagnosis also marks a moment in which pain experts and scientists began to truly grapple with the subjectivity of pain and to search for the best measures that could capture pain's multidimensionality. Among the measures that were created and remain widely used in pain management are the NRS and VAS, in which pain is represented, on the one hand, by numbers ranging from 0 to 10 and, on the other, by a 10-cm scale on which different measurements represent increasing levels of discomfort. The use of these two scales places patient participation at the center of pain diagnoses. Chronic pain, moreover, opened the door to long-term opioid use precisely through the way in which it was and is measured. This is partially because chronic pain, a long-term condition, is assessed in the short term, often in the brief span of time that constitutes a clinical appointment. In this small window of time, the practices and instruments that make it possible to attend to the subjectivity of pain (by relying on patient self-reports) also render a chronic, idiosyncratic condition as an immediate, objective problem that lends itself to similarly immediate, technological solutions—namely, opioids.

As will be discussed in the next section, the immediacy of chronic pain, when combined with the moral imperative to relieve the social problem of human suffering, helped to normalize opioids as both the gold standard treatment in pain management and the social responsibility of all physicians who treat it.

The Morality of Cancer Pain

Throughout the late 1970s and 1980s, the problem of pain was debated in cultural and political discourse related to the "War on Cancer," which helped fuel a movement of patients and patient-allies who decried the health industry's inhumane treatment of cancer patients, while promoting the recognition of patient autonomy and a moral framework around the clinical practice of cancer-related pain.

The problem of cancer pain first stepped into the limelight in the 1980s, at about the same time as the meaning of cancer itself began to shift. In these years, as cancer treatments evolved and became increasingly available, it likewise became clear that cancer was no longer a death sentence but should instead be seen as a chronic and not necessarily terminal condition. Seen in this way, cancer treatment also began to consider the question of how to treat people who must live with cancer and who, in many cases, must also live with cancer-related pain. As cancer pain became a new object of concern for many cancer patients and their doctors, the argument that this condition was not yet being adequately treated evolved as a key debate within the field. Cancer pain undertreatment thus became a hotbed for pain research, one that provided a basis for the intensified development of new treatments and expanded the clinical problem of pain and its management into the broader realm of global health.

One key event that marks this expansion is the development of WHO's 1986 report, "Cancer Pain Relief." The report itself is significant not only in that it signals a growing awareness of the problem of pain and the necessity of efforts to improve its medical management, but also in the ways in which it contributed to the narrowing of the field of pain medicine and its clinical practice. Numerous scholars and practitioners of pain medicine have commented on the significance of this report (e.g., Ballantyne & Sullivan, 2015; Norton et al., 1999; Seymour et al., 2005); in an article published in the prestigious *Journal of Pain and Symptom Management* (2005), Marcia Meldrum noted that soon after its development "hardly any article discussing cancer pain did not make reference to the WHO method" (p. 51). Doubtless, the political clout of WHO had an effect in bolstering the report's authority and its ability to shape the field of pain medicine in a meaningful way. And the report—along with the method it developed for the management of pain—is credited with having fixed an important standard for dealing with pain in both medical practice and in national drug policies.

Importantly, WHO's initiative to improve pain management also marks an early instance in which the medical problem of providing adequate pain relief was articulated within a framework that was both scientific and moral, in which a medical concept of pain relief was directly articulated to a rights-based discourse of freedom.

The WHO report was developed over several years, and the method it espouses was influenced by medical expertise emanating from two distinct traditions in pain management. The first was represented by eminent cancer pain specialist Kathleen Foley, who was stationed at Memorial Sloan-Kettering Center in New York, where pain specialists were conducting analgesic trials with cancer patients. The Sloan-Kettering approach to pain management was focused mainly on examining the effects of analgesic drugs and identifying the precise differences between and among them. The second was rooted in the British hospice care movement and a philosophy of pain management that prioritized the patient's spiritual, psychological, and physical well-being (Meldrum, 2003, 2005; Saunders 1959, 1978). While these two groups varied in their objectives—the search for an ideal analgesic, on the one hand, and the achievement of patient well-being, on the other—both advocated for a patient-centered approach to pain management that included the regular administration of pain-relieving medication. So too does the WHO's report state that "The use of analgesic drugs is the mainstay of cancer pain management" (WHO 1986, p. 18).

In its recommendations related to the practice of drug therapy, the report recommends a specific method for achieving cancer pain relief, which involves the use of a "3-step analgesic ladder" that visualizes a treatment protocol based on the premise that doctors and health practitioners "should learn to use a few drugs well" (p. 18). That the protocol is visualized as a ladder presupposes the idea of titration in drug therapy, which assumes that the proper dose is the one that provides relief and therefore involves increasing a patient's dosage step by step, until they no longer feel pain.

When looking at images of WHO's pain ladder (see Figure 1.1 below), it is easy to identify the connection between pain, opioids, and human rights: The floor beneath the ladder indicates the patient's experience of "pain," where all pain management must begin. From there, the patient starts climbing a wall that leads to the first step. The initial climb is accomplished with the use of "non-opioids," (a weak analgesic such as aspirin). The patient who makes the "non-opioid" climb and reaches the second step, which indicates the experience of "pain persisting or increasing," will then begin to climb again, this

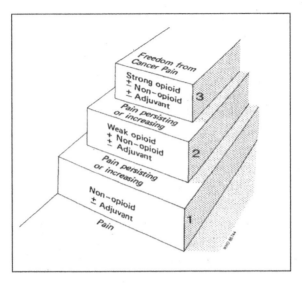

Figure 1.1. The World Health Organization's 1986 three-step analgesic "ladder"
Note: Adapted from CDP Connect (https://www.cpdconnect.nhs.scot/media/1698/controlofpaininadultswithcancer-11-appendices.pdf)

time with the help of "weak opioids." Continuing upward is a third step/state of "pain persisting or increasing" which the patient can supersede by climbing the next wall with support from "strong opioids." Finally, the "strong opioid" climb leads to the top of the ladder—the final state of "freedom from cancer pain." This analogy between pain relief and freedom establishes an important connection between pain management and human rights, one that would later be strengthened in health-related law and policy. For what WHO's analgesic ladder communicated in visual terms was soon after re-emphasized in a speech by Jan Stjernsward, the chief of WHO's Cancer Unit, who declared that "if the people know that freedom from cancer pain is their right, I am sure they will demand it. That will certainly change things" (132) (quoted in Dahl, 1993). Additionally, the text of WHO's report called upon governments to address the problem of chronic pain management within their citizenries. Those countries that recognized the authority of WHO were encouraged to think about the problem of pain undertreatment in human rights-based terms, which could be addressed through a variety of means, including government-funded pain research and liberalized national drug policies.

The significance of the WHO report also lies in its legitimation of the use of opioids for pain as a therapeutic skill rather than a risky practice. That is, that an agency charged with the oversight of international public health would publish a document declaring opioids to be the mainstay of pain management is noteworthy in and of itself. WHO's paradigmatic "analgesic ladder" and its widespread uptake in clinical practice constructs a clinical paradigm in which pain and opioids are bound up together, where opioids are represented as a sure solution to the clinical problem of pain and the moral dilemma of assuring patients' right to freedom from it.

Soon after WHO published its report on cancer pain relief, many states began adopting "intractable pain regulations," which sought to confront the problem of chronic pain undertreatment—which was by then becoming more widely recognized as a growing social problem. The regulations provide a concrete example of how the clinical practice of pain management is governed at the intersection of medicine and politics, where it is delimited by morally inflected discourses and discussions of how pain should be treated and to what extent physicians should be deemed responsible for providing it. The "Intractable Pain Treatment Acts" (IPTAs) that many states soon adopted (beginning with Texas in 1989) prohibited the persecution of physicians who prescribed opioids for pain, so long as their practice could be considered within the bounds of responsible healthcare. Such a policy was meant to encourage physicians to utilize narcotics when treating chronic pain—a practice that, as discussed, many physicians had previously avoided, both because of their fears that patients would become addicted and their concerns that such events would result in their being sued for malpractice. It is no surprise then, that the movement for states to adopt IPTAs was often led by physicians who were trying to respond to the cultural stigmas and restrictive regulations that they saw as preventing them from prescribing opioids to their chronic pain patients. Yet, somewhat paradoxically, the IPTAs also had the effect of instigating their own wave of malpractice lawsuits against physicians, namely, by patients who felt their pain had not been adequately treated and warranted an opioid prescription. Thus, while IPTAs granted physicians more freedom in terms of their ability to prescribe opioids, these freedoms were also limited by the possibility that physicians who refused to do so could now be punished.

IPTAs mark a concrete example of the ethical questions that have arisen in pain management, and which situate pain and opioids within a distinctly moral domain, where opioids have become understood as part of a physician's responsibility to manage patients' pain. In recent years, these questions have crystallized in ongoing debates and court cases waged around the associated problems of pain undertreatment and—in the wake of a mounting crisis—its overtreatment. These issues, and the crucial problem to which they give rise, have laid the groundwork for transformations in opioid prescribing that undergird the present crisis.

In 2016, Dr. Hsin-Ying "Lisa" Tseng became the first doctor convicted of murder for recklessly prescribing drugs. The Southern California–based physician, operating out of her office in a strip mall, was charged with the overdose deaths of three of her patients and sentenced to 30 years to life in prison. While Dr. Tseng's case marked a precedent for physicians being charged with the murders of the patients to whom they prescribed painkillers, it was but one among several cases that developed in the same years around the country, in which physicians were repositioned as "doctors-turned-drug dealers" (Girion & Glover, 2016). Some of these cases involved health professionals participating in clearly illegal prescribing schemes, namely, by accepting positions on the payroll of criminal "pill mills"—pain clinic operations where doctors knowingly write prescriptions without a legitimate purpose for any patient who is willing to pay for them. Others, however, have involved self-employed or hospital-employed individuals, whose motives for prescribing dangerous quantities of powerful drugs have not always been clear.

The case of Dr. William Husel, an Ohio-based acute care doctor charged with the murders of 25 of his patients, is emblematic of a case in which the prescriber's intentions—and, by extension, the intent behind their crimes—are difficult to surmise. Dr. Husel's trial, one of the biggest-ever murder cases against a healthcare professional, concerned the overdose deaths of the more than two dozen patients to whom he prescribed massive amounts of painkillers over the course of three years. The dosages he mandated for those in his care, several of whom were terminal, end-of-life patients, were thought to have hastened or caused their deaths. The quantities were suspicious; in some cases, they overrode the hospital's drug protocols and, in others, were declared to have been anywhere from 10 to 40 times higher than the amount considered to be medically appropriate for treating chronic pain.

Some of the coverage of Dr. Husel's and Dr. Tseng's cases characterized them as exemplifying a situation where "dirty doctors" transformed into drug dealers, namely, by prescribing quantities that ensured their patients would remain addicted to their medications and, in doing so, guaranteed ongoing income for the doctors themselves (Hassan, 2019). But the doctors, in their defenses, claimed that this was not the case, and that what they really had in mind when they wrote their patients' opioid prescriptions was not profit, but human suffering—in particular, the intense daily suffering of chronic pain patients, which they felt responsible for mitigating. In Husel's case, the questions of how to alleviate the suffering of his terminal patients and what kind of dosing regimen should be implemented in order to do so seem to have been significant ones, at least among the various nurses, pharmacists, and managers working under him who, despite any doubts they might have had, were satisfied enough by the explanations he provided to administer or otherwise provide the amounts of medication that Husel recommended. While none of these individuals were accused of any crime or charged with having been complicit in the treatment outcomes, dozens were eventually placed on administrative leave, fired, or suspended in the face of pending action from nursing agencies. Their involvement in the case is suggestive of a lingering doubt regarding who, exactly, was culpable in the case of these 25 overdoses and to what extent they should have been held responsible.

The key question these cases raise is that if pain is to be seen as a human rights issue, and if physicians are obligated to protect and provide this right to patients, then what is the extent to which they can—or must—go in order to do so? How are lawmakers, physicians, and other health professionals to measure human suffering? What metric can be used that will prove adequate to enable the development of clear limits for defining the appropriate extent to which chronic pain can and should be treated? No concrete answers have been developed to address any of these crucial questions. They belong to the hazy knowledge terrain and uncertainties that have long haunted pain management—uncertainties that have been marked first and foremost by the problem of subjective pain.

Subjective Pain and the Fifth Vital Sign

The moral quandaries that frame the issue of pain management in global health, law, and politics have been made increasingly complicated by widespread uncertainties regarding what pain is and the subjective nature of it. These uncertainties, moreover, have paved the way for political discussions that would ultimately dictate the ways in which pain is to be understood, measured, and intervened upon. In turn, these decisions have laid the foundations for the current crisis.

While the enactment of IPTAs laid down a legal precedent for understanding pain relief as a human right, they did not answer the question of how the highly subjective experience of pain, on the one hand, and relief, on the other, should be assessed and measured. The measurement of subjective pain—which was acknowledged by John Bonica early in the field's development and, after years of debate within the field, has since become the basis of pain's medical definition—represents another important conjuncture in the genealogy of pain management. At this conjuncture, the moral obligation to provide citizens with their right to pain relief is complicated by a larger political struggle around the best way to properly assess pain's subjectivity.

As Keith Wailoo (2015) has written, the subjective nature of pain has long posed a problem for politics. It has formed a key battleground between conservatives and progressives in debates about welfare, the former of whom have been (and remain) concerned about disabled "malingerers," those who fake their pain in order to secure disability benefits. In the Reagan years, conservatives attempted to impose rigorous disability tests on the assumption that many of the disabled—particularly those who claimed to suffer from chronic pain—where faking or exaggerating their symptoms. Dismissing progressive pain standards, which largely accepted individual's self-described pain symptoms as sufficient evidence of a disability, Reagan's administration would dramatically reform the country's welfare system in order to distinguish between those whose pain was "real" and those whose pain was merely "in their heads." The conservative formula for judging people's pain was first addressed in the Social Security Disability Benefits Reform Act (henceforth referred to as the "Reform Act") of 1984. The Reform Act was enacted as a response to the Social Security Amendments of 1980, which were passed in order to temper the skyrocketing cost of supplemental security income (SSI) and social security disability insurance (SSDI). The main change enacted by the 1980 amendments was a new requirement that recipients participate in an eligibility review process every three years. Yet, this attempt to weed out false disability claims ended up resulting in hundreds of thousands of recipients having their benefits terminated (Rodgers, 1991). Four years later, the Reform Act was developed in response to widespread public outcry and demands that deserving recipients (namely, mothers and soldiers) have their benefits reinstated (Skocpol, 1995). In response, Congress passed the Reform Act, which put the burden of proof on disability decision makers to provide objective evidence that a claimant's subjectively defined disability had improved before their benefits could be suspended.

For our purposes, the Reform Act's most significant feature was its Section 3, which demonstrated Congress's concern over the subjective nature of pain: During congressional deliberations, several members expressed concern over recent federal court decisions that had ruled that subjective pain testimony could be considered sufficient as evidence of a claimant's disability. These court opinions, members suggested, had moved beyond the intentions laid down in previous disability policies. Furthermore, they gave far too much weight to subjective reports of pain and thus failed to reflect the administration's current position on the evaluation of pain in disability cases. In response to these court decisions, many members of Congress began to advocate for a more restrictive definition of pain and a more rigorous process for its evaluation. Their efforts to do so bore fruit, resulting in the first statutory standard for the evaluation of pain, which states that,

> An individual's statement as to pain or other symptoms shall not alone be conclusive evidence of disability as defined in this section; there must be medical signs and findings established by medically acceptable clinical or laboratory diagnostic techniques, which show the existence of a medical impairment that results from anatomical, physiological, or psychological abnormalities which could reasonably be expected to produce the pain or other symptoms.

The evidence that would be required, moreover, was defined as, "objective medical evidence of pain or other symptoms established by medically acceptable clinical or laboratory techniques (e.g., deteriorating nerve or muscle tissue) must be considered in reaching a conclusion as to whether the individual is under a disability" (Social Security Disability Benefits Reform Act of 1984, 1799–1800).

Somewhat paradoxically then, the Reform Act acknowledged the subjective nature of pain but at the same time set a standard that would fail to cover many chronic pain sufferers (those without "objective medial evidence" of an "underlying medical impairment") under the current law. To resolve this paradox, the Reform Act then called upon the authority of medical, legal, and administrative expertise to resolve this problem by directing the Secretary of Health and Human Services to appoint an expert "Commission on the Evaluation of Pain." The Commission, as the Act lays out, would be required to conduct a study in collaboration with the National Academy of Sciences on the assessment of subjective pain and to set a medical standard for determining when an individual claiming chronic pain could be deemed eligible for disability benefits. It would contain at least 12 members from the fields of medicine, law, and disability program administration, who would then work in collaboration to address concerns about the use of subjective evidence in determining disability eligibility and to submit its evaluation to congressional committees the following year.

Thus, in the 1984 Reform Act, Congress intervened in the political problem of pain not by eliminating the possibility for pain to be recognized as a disability, but through legislation that restricted the definition of pain to "objective medical evidence" and other signs and symptoms established by "medically acceptable clinical or laboratory techniques" (Social Security Disability Benefits Reform Act of 1984, 1799–1800). The amendment's reliance on the authority of expertise—as manifested in its call for more science, objectivity, and neutrality—set a political precedent for the medical practice of managing pain based primarily on the existence of observable evidence. In doing so, the policy also hints at the triumph of evidence-based medicine (EBM) in science and politics—an epistemological standard that would, alongside the Reform Act, call for objective evidence as the "gold standard" for defining pain and relief from the 1980s onward.

The development and normalization of EBM has limited the practice of pain management in clinical contexts. EBM was first defined by healthcare analyst, mathematician, and physician David M. Eddy and was consolidated in the second half of the 1980s in a series of population-level policies, guidelines, and reports related to healthcare decision-making. It has since become the key standard for decision-making across nearly all levels of healthcare—in the development of health-related policies, the design and implementation of medical research, and, eventually, in clinical practice (see Eddy 1982, 1984, 1988; Beltran, 2005). In brief, EBM refers to an approach to medical practice that emphasizes the use of evidence from research, typically conducted in randomized control trials. According to this approach, the quality of "evidence" depends on its epistemological strength, which is associated with research methodologies that attempt to control for human bias (e.g., the placebo effect). Today, EBM guides not only medical research but also decision-making at nearly every level of our healthcare system.

EBM and debates around social welfare came together and crystallized in the development of clinical guidelines that define and measure pain as a "vital sign" and, in doing so, resolve the scientific and moral problem of pain treatment. The treatment of pain as a vital sign, equivalent in importance to a patient's heart rate, temperature, breathing rate, and so on, illustrates just how central the measurement of pain had become to medicine by the end of the century and how the objective measurement of subjective experience became the go-to method for assessing pain and pain relief across the country—a tendency that has accompanied upsurges in opioid prescribing since the end of the 20th century.

In 1996, the president of the American Pain Society, Dr. James Campbell, gave a speech in which he stated that "if pain were assessed with the same zeal as other signs (vital) are, it would have a much better chance of being treated properly. We need to train doctors and nurses to treat pain as a vital sign" (Campbell, 1996). Three years later, the U.S. Department of Veterans Affairs (DVA) added pain as a "fifth vital sign" to be regularly assessed as a major indicator of health. As the *Washington Post* reported, "VA officials said the change in routine is designed to call physicians' attention to what is widely considered one of the most unrecognized and untreated symptoms in American health care" (Schuster, 1999).

In the years that followed, Campbell's suggestion that pain be recognized as a sign in equal importance to other vital signs, and that it be consistently and regularly measured in the clinical context, was gradually institutionalized, first in the Veterans Health Administration, which published a report in 1998 declaring the implementation of a "comprehensive national strategy for pain management" in which pain would be treated like a vital sign—in order to "trigger further assessment of the pain, prompt intervention, and follow-up evaluation"—and then across clinics and hospitals across the nation (DVA, 2000, p. 5). Soon after the VHA published its report and "toolkit" for the implementation of its new pain management strategy, the Joint Commission on Accreditation of Health Care Organizations (JCAHO) followed suit by releasing a set of new pain management standards that echoed the sentiments being expressed by the VA, American Pain Society, and other professional societies dedicated to spreading awareness of the problem of untreated and undertreated chronic pain. The Joint Commission's standards declared that all "patients have the right to appropriate assessment and management of pain" (JCAHO, 2001, p. 2) and laid out several tools that practitioners could use in order to ensure that their patients' rights were acknowledged and protected. JCAHO's new pain management standards were likewise accompanied by a wave of clinical practice guidelines for pain management, which emerged from a variety of advocacy groups, professional societies, and other healthcare organizations, including the American Geriatric Society

(1998), American Medical Directors Association (1999), U.S. Department of Health and Human Services' Agency for Health Care Policy and Research (Jacox et al., 1994), the American Academy of Family Physicians (Marcus, 2000), and the American Society of Anesthesiologists (Wilson et al., 1997).

Common to all these guidelines is the mandate that this new vital sign be assessed using an NRS, and that this assessment must be performed in every clinical encounter. Measuring pain as a vital sign can be seen as an institutional response to the need for "more objectivity" in pain management and for the design of tools that enable the objective measurement of pain in terms of its level of intensity and, after treatment, of relief.

There is a contradiction inherent to the VA's guidelines, which attempt to assess and manage a multidimensional experience using unidimensional tools, even while the organization recognizes that pain is a subjective phenomenon and that patients' self-reports are "the single most reliable indicator of pain," followed by reports from patients' family and friends (DVA, 2000, p. 13). However, it also seems that the use of measurements such as the NRS go some way toward resolving this contradiction: A patient's assertion that his pain is a level 10 is, after all, a kind of self-report, but one that has been abstracted from lived experience. Through tools such as the NRS, which reduce pain to a number, the complexity of its lived experience is made manageable. So too does this abstraction attempt to address the problem of pain's psychological dimensions. For there is an additional struggle in assessing illnesses that, like pain, are often lacking physiological markers. In the measurement of mental illnesses such as depression, as in the case of measuring complicated chronic pain, the absence of visible signs of illness (whether physiological or behavioral) shores up a need for tools (such as NRSs) that promote consistency and attempt, as Andrew Lakoff has written, to "turn amorphous, heterogeneous experience into a calculable problem" and, in doing so, to limit the bias of evaluators' subjective judgment (Lakoff, 2007, p. 58).

And there are other practical reasons underlying the use of NRSs for measuring pain, which have to do with its clinical context: One has only to consider the busy, sometimes chaotic environment of most clinics, where tools such as the NRS offer doctors and nurses a means of rapidly assessing, treating, and reassessing pain and pain interventions (Pasero & McCaffery, 1997). In such cases, unidimensional measurements offer many benefits: They enable practitioners to take patient experience into account within the limits of a 5-to-10-minute-long appointment, while also transforming this experience into a number. In doing so, such measurements help create a universal vocabulary for the clinical practice of treating patients' pain. Moreover, they can be useful tools for justifying the use of particular protocols or practices to outside observers who are concerned with the effects that human bias may have on the diagnostic process.

The idea that pain interventions should quantitatively demonstrate the effects of a drug and map them directly onto a specific illness is supported by the assumption that diseases are stable entities that can be explained by simple causal mechanisms. This idea, which underlies the use of NRSs in pain management, is also consistent with the logic of specificity that underlies drug development (Lakoff, 2007). That is, drugs purport to act directly on disease in a way that other kinds of interventions (behavioral therapies, psychoanalysis, physical therapy, occupational training, etc.) do not.

Crucial to the story of opioids is the fact that the NRS, which reinforces the logic of specificity in drug development, also supports a theory of pain management where pain is defined in terms of pain-relieving drugs. Put another way, the numerical scale for measuring pain measures two variables: the intensity of the pain itself and the degree of relief, which is measured after a treatment has been administered. What this means is that the scale does not only measure pain. It is also designed to measure the efficacy of pain medications. This is particularly significant when considering the other key component of clinical practice guidelines discussed in this chapter—the mandatory scheduling of routine follow-up appointments for all pain patients—since the first appointment in which a patient's pain is first assessed and diagnosed anticipates the possibility of future diagnoses and the ongoing use of medication.

The VA's movement to address pain as the "fifth vital sign," which was quickly adopted by other organizations within the medical community, was also a movement to redefine pain—to cast what otherwise would have been understood as a purely subjective phenomenon in objective terms, as something that could be accurately measured and quantified. Crucially, these new tools for objectively measuring the intensity of pain worked to separate the patient from their pain, abstracting the phenomenology of their lived experience and rendering it as a quantified variable and medical/scientific fact. Quantified pain thus represents a radical departure from the consensus within pain medicine's multidisciplinary community (and which had been established in John Bonica's 1953 textbook) that defined pain as a multidimensional and subjective phenomenon, which must be understood primarily in terms of its lived experience—its phenomenology, the ways in which people in pain experience and respond to it. In clinical research and practice, quantified pain legitimates and normalizes (and is, in turn, legitimated and normalized by) the focus on opioid medications as the mainstay of treatment. For if, upon reassessment, a patient reports a lower level of pain, their self-report is viewed not as a change in perception but as a correction—proof that the treatment is working and should continue to be administered. A treatment that is held to be effective, moreover, is seen as directly acting on pain itself, rather than on perception or on the environmental or contextual factors that may also influence how patients experience their pain.

The development of the guidelines promoting the measurement of pain as a "fifth vital sign" can therefore be understood as a moment that foreshadowed the ongoing opioid crisis. The moment at which pain became meaningful as a vital sign is crucial, but not only because it expanded the recognition and measurement of pain in clinical contexts. It also transformed the definition of pain and repositioned it in a context that changed its nature by attaching it to government initiatives and to a rapidly transforming landscape of healthcare and medicine—one dominated by the rules, norms, and values of evidence-based medicine and biomedical models.

* * *

In October 2000, Congress passed the Conquering Pain Act of 2001, a bipartisan effort. The Act declared the beginning of a "Decade of Pain Control and Research"—a new medical era that would be characterized by an intensified focus on promoting "research, education, and clinical practice" (American Medical Association House of Delegates, 2000). The Conquering Pain Act provides a final example of the ways in which the U.S. government prioritizes "objective" science as the most viable method of ensuring its citizens' right to pain relief. The Conquering Pain Act's proposal to address the problem of chronic pain through the development of more objective, quantitative evidence reinforces the assumption that the right to relief is best secured through the sponsorship of research that adheres to the standards of EBM. Such an assumption, which values neutrality and objectivity, also obscures the reality that pain is anything but those things. It ignores an important reality that pain is unevenly distributed in society and that patients' access to relief is not only limited because of a lack of research/education but is also restricted by structural constraints. Without acknowledging the other conditions such as institutional norms, cultural stigmas, and systemic inequalities that underlie the nation's high incidence in undertreated pain, we are destined to remain tethered to a limited idea of what good pain management is and could be—one that, for the moment at least, is highly dependent on unidimensional measures and pharmacological solutions. And, as has been discussed in this chapter, these are the same solutions that can be understood as having anticipated many of the problems we now face as the nationwide rates of chronic pain and opioid-related overdose continued their ascent.

2
Strategic Ignorance in Opioid Regulation

You could say that for the U.S. Food and Drug Administration (FDA), the story of modern opioids began in December 1995, when the then-small, family-owned Purdue Pharma received a letter from them, approving their application for a new prescription painkiller—a promising opioid named OxyContin—for the management of moderate to severe pain (FDA OxyContin NDA, 1995). Immediately, the company began promoting its drug widely. Its efforts quickly bore fruit: OxyContin prescriptions for noncancer pain patients increased tenfold, jumping from 670,000 in 1997 to 6.2 million in 2002, while prescriptions for cancer-related pain increased fourfold, from 250,000 to 1 million (Van Zee, 2009). Meanwhile, dozens of stories began surfacing in local and regional newspapers that warned readers of a dangerous new painkiller and a slew of problems associated with it: Already by 1999, the DEA reported that 85% of arrests for false prescriptions in Maryland were written for OxyContin and other oxycodone products. The following year, the Office of National Drug Control Policy stated that OxyContin abuse had emerged as a significant problem in Baltimore, Boston, Billings, Denver, Detroit, Honolulu, New Orleans, Philadelphia, St. Louis, and Washington, DC (Diversion Control Program, DEA, 2001). In April 2002, the DEA conducted an expanded review of autopsy data to assess the damage that OxyContin appeared to be wreaking on local communities and concluded that the drug had played a role in at least 464 overdose deaths. In all but 10 of the overdoses, people who use opioids had taken the drug orally—just how it had been prescribed (DOJ, 2001).

The rapid uptake of OxyContin and the rise of overdoses and deaths linked to it clearly showed that this drug, contrary to the FDA's initial assessment of it, is dangerous. Why, then, did the FDA approve it? And why, some 20 years later, has it continued to approve new formulations of OxyContin and similarly potent opioids? Long before the FDA first approved opioids for medical use, it was already widely known that such drugs are habit forming and present a significant risk of accidental overdose. But this knowledge in discussions about opioid addiction spans back centuries. Was this history unfamiliar to regulators? Did they know? Or were they ignorant?

We can't answer these questions without looking more closely at the process of opioid regulation and the logics that underlie it. In this chapter, I review the regulatory process of OxyContin to trace the steps that the FDA took to evaluate the safety and efficacy of Purdue Pharma's most famous painkiller. I also look carefully at the postmarket context of regulation, where the story of OxyContin becomes particularly interesting—following the first reports of its abuse and the early days when these reports began piling up, prompting a series of postmarket regulatory actions. Through an analysis of FDA advisory meetings, congressional hearings, regulatory planning documents, and other sources, I outline what I believe are the most crucial moments in the regulatory history of the opioid crisis.

Combing through FDA documents, it becomes clear that the fulcrum around which the story of opioid regulation pivots is also pain. Specifically, the trajectory of opioid regulation has been marked time and time again by extreme uncertainty around pain. Unknowns about how pain functions have gone hand in hand with a style of institutionally managing opioids that helped kick-start the current crisis. The logics that are responsible for making our health system function, and that can also be shown to have played a direct role in the opioid epidemic, are defined by uncertainty as well as strategy; many of the mistakes that regulators made in failing to anticipate the risks of drugs like OxyContin were made not despite the unknown, but in light of it. Decision-making in opioid regulation has been conditioned by a logic of strategic ignorance. As Linsey McGoey (2019) has written,[1] ignorance is not always the opposite of knowledge, though it is often the case that knowledge is seen as power while ignorance is seen as weakness. Like McGoey, I am suspicious of this juxtaposition. Sometimes ignorance—burying one's head in the sand—is not passive but productive, not a sign of vulnerability but a strategy. Employed strategically, ignorance also produces knowledge, although it may be limited. The knowledge that ignorance produces, moreover, has social effects. These can be mild, or they can, as in the case of an opioid crisis, be very serious.

In this chapter, we see how strategic ignorance related to the question of pain and its management was leveraged in opioid regulation and what kinds

---

[1] McGoey (2019) independently formulated a similar version of strategic ignorance in her analysis of ignorance and power in the modern age.

of social effects have followed in its wake. Strategic ignorance has worked in ways that were (and are) geared toward advancing the interests of pharmaceutical companies and the continued production of opioids. This is because it is reinforced by other regulatory logics that have long characterized pharmaceutical regulation in the United States. I refer to these regulatory logics as permissive logics because, rather than acting preventatively to anticipate possible risks, they do the opposite by assuming that if a risk cannot be clearly measured, then it doesn't exist.[2]

Permissive logics are also noninterventionist logics and, in drug regulation, go hand in hand with the expansion of pharmaceutical markets. In the opioid market, where supply and demand continue to increase, these logics have laid an extremely fertile ground out of which the current crisis could grow. The growth of new, semilegal markets for opioids is part of the market expansion that has taken place alongside the epidemic. I refer to these new systems of exchange as the opioid gray market, which have presented regulators with novel risks that continue to disrupt the regulatory process and render useless many of the tools that regulatory institutions have for dealing with the current crisis.

The Production of Strategic Ignorance

Safety and efficacy are the twin pillars of U.S. drug regulation and form the basis of the FDA's approach to assessing the viability of new drugs. Before a drug enters the market, it must be shown to be both safe and effective for its target population. While these principles may seem obvious, they were not always a part of U.S. drug regulation. The requirement for drug manufacturers to demonstrate their product's safety was enacted in 1938, with the passing of the Food, Drug, and Cosmetic Act—a piece of legislation that was developed to respond to the public harms inflicted by the patent drug industry.[3] And it wasn't until 1962, when Congress passed the Kefouver–Harris Amendments, that the FDA gained authority to require evidence of the effectiveness of drug makers' products. Though what this meant in practice was that a drug

---

[2] Nancy Langston (2010) has laid out a similar argument in her analysis of diethylstilbestrol (DES) and with regard to the FDA's logics and attitude toward mitigating the risks associated with endocrine-disrupting chemicals.

[3] It's worth recalling Mrs. Winslow's Soothing Syrup, a patent medication for soothing fussy babies, which contained a combination of morphine, alcohol, and laudanum and remained in use throughout the 1930s.

must be shown to be no less effective than previous FDA-approved drugs of the same class. It is up to the drug's industry sponsor to demonstrate that the substances it manufactures meet these requirements. Historian Daniel Carpenter has argued that defining pharmaceutical politics solely in terms of safety and efficacy is problematic, partially because this two-pronged focus ends up excluding other important questions from the regulatory process, like the heterogeneity in individual responses to drug treatment, the use of placebo effects, and the effect that drug advertisers have on consumers (Carpenter, 2010, 16–17). These questions are crucial for pain management since pain is not uniformly experienced. But they were, from the beginning, excluded from OxyContin clinical trials, which sought to measure pain.

The first clinical trials for OxyContin were conducted in 1989 among 90 women in two hospitals in Puerto Rico. The study involved only female patients, all of whom suffered acute pain from surgeries, and none of whom were characteristic of the patients OxyContin would be indicated for. To approve OxyContin's release, the FDA required that it be proven effective for at least 12 hours in at least 50% of the patients who tried it, a feasible target since surgical pain is considered easier to treat with targeted interventions than chronic pain, which involves a variety of psychological and social factors. Yet, as the *L.A. Times* reported, more than 30% of the women who were given OxyContin reported that their pain returned within the first eight hours, and nearly 50% of them reported needing more medication before the 12-hour mark (Ryan et al., 2016). This finding is compounded by the fact that unlike chronic pain sufferers, who often have experience using opioids, the women who participated in the OxyContin trials were opioid naive, with little or no experience taking opioids. As such, their tolerance was likely lower than that of the average chronic pain patient. The medication should have had a stronger impact on them, with effects that should have lasted longer than they would in people with different levels of experience taking similar drugs. And yet, fortuitously for OxyContin manufacturer Purdue Pharma, these results were enough to assert that OxyContin was not only safe but also effective (or rather no less effective) than other opioids.[4]

These results make it difficult to account for much of what defines pain, like heterogeneity, subjectivity, and social factors. But the logics of the trials themselves produce this ignorance. The logic of control that is central to their

---

[4] This status would ultimately (though misleadingly) lead Purdue to advertise its new drug as being superior to other painkillers in its class.

design requires research subjects to take a drug uniformly, in a prescribed way. For in a controlled study, subjects are not permitted to manipulate drugs or deviate from their dosing schedule. If deviations were to occur, they would generally do so outside of the trial context and as such would not be observed or recorded in study findings.

How could a study like this, formed around a homogenous group of opioid-naive research subjects and tethered to strict consumption behaviors, represent the reality of painkiller use? Of course, it could not have. The clinical trials of OxyContin are therefore useful for showing how strategic ignorance functions in opioid regulation, and it how it has done so since the beginning of the opioid epidemic: While OxyContin's clinical trials did produce knowledge about the drug, this knowledge was strategically limited. And when it became approved, much was still unknown about OxyContin and the patients whose pain Purdue Pharma would gain the right to treat.

In addition to evidence gleaned from clinical trials, other data was used to prove OxyContin's safety and efficacy. This data was lifted from recent history in the form of a morphine-based painkiller named MS Contin, which had been approved by the FDA in 1987 and was also developed by Purdue Pharma. Like OxyContin, Purdue's MS Contin offered a novel benefit for people who use opioids. It had a slow-release mechanism, which would lengthen the pain-relieving effects of its morphine for 12 hours. Thus, MS Contin would require consumers to dose only two times a day rather than every four to six hours, as was the case with every other opioid medication on the market at the time.

In recent years both the FDA and Purdue Pharma have frequently responded to accusations blaming them for the opioid epidemic by pointing to MS Contin, its safety record, and the similarities between it and OxyContin that led them to believe the latter would be just as safe and even more effective than its predecessor. Their argument is that because MS Contin was proven safe and because OxyContin so closely resembles this older drug in both form and function, no one could have anticipated the negative consequences of the newer drug. OxyContin presented no foreseeable risks in its history or its research. Clinically and historically, it appeared safe and effective, so its release onto the market was a logical choice.

But in OxyContin's early years on the market, it was already becoming clear that the drug posed several risks to its consumers. Congress convened a series of hearings during which FDA officials, industry representatives, and other stakeholders were called upon to evaluate the drug's risks and benefits and to assess the Administration's rationale for approving it. In two separate hearings in 2002 and 2005, the FDA's then Director of the Center for Drug Evaluation and Research, Robert Meyer, and its then Director of the Office of New Drugs (OND), John K. Jenkins, offered statements explaining its lack of foresight, and its boiler plate explanation. Based on the information available to the FDA, including "the record of other modified release Schedule II opioids," the nationwide "abuse and misuse of OxyContin reported over the past few years were not predicted." The FDA believed that the controlled-release mechanisms that make the OxyContin formulation last longer would also result in less abuse potential, since the drug "would be absorbed slowly and there would not be an immediate 'rush' or high that would promote abuse" (FDA, n.d.; Jenkins, 2002, p. 14; U.S. House Committee on Government Reform, 2005, p. 30). Moreover, as Director Meyer emphasized in a 2005 hearing before the House Committee on Government Reform, the FDA had no reason to assume that OxyContin's risks would outweigh its benefits because MS Contin had been safely marketed in the United States for years (Meyer, 2005, p. 30), though MS Contin never had the spotless safety record that Meyer claimed it had. Three years after it hit the market, a *Cancer* article demonstrated skyrocketing rates of MS Contin abuse in Ohio, particularly in the area in and around Cincinnati. According to the authors, shortly after it was approved, MS Contin also became a "highly desirable preparation for opioid abuse" (Crews & Denson, 1990, p. 2642). The drug had even surpassed the popular painkiller hydrocodone, or Vicodin, as the most abused prescription opioid in the area. Apparently, a huge quantity of morphine could be extracted from Purdue's 12-hour pills, which could also be injected intravenously for a long-lasting effect (Crews & Denson, 1990). But the FDA was caught off guard when, some 10 years later, the same thing happened with OxyContin. For in theory, OxyContin offered the same mechanism but with oxycodone instead of morphine—a substance calculated to be around 1.5 times stronger than its morphine-based antecedent (Stanford School of Medicine Palliative Care, 2017). When the pill was not swallowed but instead was crushed or chewed or melted down into a fluid, it transformed into an instant-release medication, one that would provide a stronger effect than MS Contin.

The FDA defines its approach to risk management as "safety or risk assessment combined with efforts to minimize *known risks*" (Meyer, 2005, p. 35, emphasis added). One of the many problems inherent to this definition of risk management, which stands out in the case of the Administration's quick approval of OxyContin, is its reliance on the "known." Trusting what you already know to help you anticipate a future that can only be unknown is problematic for a number of reasons. For the FDA, doing so has meant relying on limited data and not thinking through possible risk scenarios associated with OxyContin, which rendered it blind to the future.

This approach to risk management also exemplifies a rejection of what others have identified as the "precautionary principle in drug regulation" (Langston, 2010; Sunstein, 2002). As Cass Sunstein explains in his analysis of risk and reason, a precautionary approach assumes that it is "better to be safe than sorry" and, in doing so, takes uncertain risks into consideration.

The FDA's risk management framework might be better understood as having adhered to a permissive risk logic, instead of a precautionary one. A permissive approach to risk does not consider risks that are not already known. Instead, it assumes that what is unknown does not exist (Langston, 2010). This is exemplified in the idea that so long as a drug is considered safe and effective, or so long as it is no less safe and no less effective than other FDA-approved drugs of the same class, then its potential risks are outweighed by its potential benefits. When regulation operates according to this principle, it does not place the burden of proof on drug sponsors to prove their drug's worth but, rather, tends toward the approval of products, in this case even when they don't adhere to the FDA's twin pillars. Permissive logics were also at work when the FDA's medical review officer signed off on the New Drug Application (NDA) for OxyContin, even though that officer had concluded earlier in the document that the efficacy of the drug was *not* higher than other drugs on the market but was only "equivalent" to them (Purdue Pharma, 1995, n.p.). Yet that is when OxyContin first entered the pharmaceutical market.

When it comes to minimizing potential risks at the FDA, risk communication is defined as the "cornerstone of risk management efforts for prescription drugs" (FDA, 2005a). This being the case, one of the first preventative measures the FDA took to mitigate OxyContin abuse was to include a black box warning on all OxyContin packages informing patients that "OxyContin tablets are to be swallowed whole, and are not to be broken, chewed, or crushed. Taking broken, chewed, or crushed OxyContin Tablets could lead to the rapid release and absorption of a potentially toxic dose of oxycodone" (OxyContin, 1995). The thick-framed "black box" on a product's packaging constitutes the highest level of warning for an approved product and one of the more emphatic risk communication strategies the agency uses.

The FDA's decision to include a black box warning was made in light of the knowledge that the drug about to be approved could be manipulated in ways that would defeat its controlled-release mechanism. This knowledge was published in Purdue Pharma's toxicology reports, which showed that 68% of the pure oxycodone in the pill could be easily extracted either by crushing the pill or dissolving it in liquid (Meyer, 2005). This fact contradicted the company's claim that because of its timed-release ability, its new product was inherently less abuse prone than other opioids on the market. The black box labeling was included as a means of minimizing this potential risk. But it failed. Rather than preventing patients from "accidentally" misusing their medication, the labeling seems to have encouraged it. OxyContin abuse and overdose incidents continued to climb. As one study put it, the FDA's black box seemed to function less as a warning and more as an advertisement that reached many consumers (Griffin III & Spillane, 2012).

The problematic logic behind the FDA's labeling decision is linked to the FDA's approach to risk minimization and to the idea that by communicating risks, one can successfully manage them. What is not taken into consideration in such an approach to communication is the possibility that while the message may be received, its reception may differ depending on the ways in which it is interpreted. Its interpretation by different kinds of receivers with differing needs conditioned the function of that message and the effects it had been demonstrated to have on the beliefs and behaviors of those who receive it.

During the same years in which OxyContin was undergoing its approval process, several studies were published that demonstrated a significant risk of abuse and "aberrant drug-related behavior" among chronic pain patients (e.g., Fishbain et al., 1992; Hoffmann, Olofsson, et al., 1995; Kouyanou et al., 1997). That pain patients do sometimes take their drugs in ways other than "as prescribed" is an important reality that the FDA did not consider. For those who suffer from chronic pain, what is considered an "effective" drug in the cleaner context of a controlled trial with naive patients may not be as effective in real life. These patients live in a more complicated reality of pain, one in which self-medication is sometimes necessary to be able to live and work without constant discomfort.

Because the FDA failed to consider the complex reality of chronic pain and painkiller use among those who live with it, it could not foresee the ways in which its risk communication strategies aligned with the reality of a heterogeneous population of consumers. Its actions suggest that the FDA assumed that the needs of OxyContin consumers would align with its own understanding of pain produced in clinical trials and that all consumers take their medication as they do in controlled contexts. The assumption that patients should—and do—take their opioids "as prescribed" is a problematic one, for it does not take into account the dynamics of pain as a lived experience.

The effect that ignorance has had in this case has been a widespread "boomerang effect," as initially described by Ulrich Beck. In *Risk Society* (1992), Beck describes how the management of risks that aren't easily measured or foreseen return to wreak havoc on the actors or institutions that aimed to mitigate them. The FDA's management of the risks of painkiller use shows us how relying on the "known" results in a permissive risk logic and how this logic distanced the FDA from the realities of chronic pain. Though once stories began surfacing in newspapers and on television about pain sufferers "misusing" or "abusing" OxyContin, the FDA's distance from the reality of pain quickly diminished.

Partial Knowledge in Opioid Pharmacovigilance

After the FDA has approved a new drug, its ability to regulate that product changes. As Carpenter (2010) explains, the moment in which the FDA signs off on an NDA can be understood as the moment "when the Administration has relinquished its gatekeeping power over a drug" (p. 586). In this late stage of regulation, the tools the FDA has at its disposal to regulate a drug are feeble at best. The best it can do is monitor the drug's consumption trends, which it does by drawing from a variety of databases such as the DEA's DAWN database, the Substance Abuse and Mental Health Services Administration's (SAMSHA's) National Survey on Drug Use and Health, and state prescription drug monitoring programs (in cases where they exist) and collect and analyze the data therein—and urge pharmaceutical companies to do the same.

The FDA refers to this practice as "pharmacovigilance," a term that encompasses the agency's efforts to "ask companies to collect special information to improve the speed and sensitivity of detecting suspected safety problems" (Meyer, 2005, p. 35). The FDA must "ask" companies to conduct pharmacovigilance efforts because it does not have the power to require them to create surveillance programs for any of their products. In fact, the FDA's authority in the postmarket regulatory realm extends only so far as it can provide companies with guidance and recommendations as to how they might plan and develop their pharmacovigilance toolset. Moreover, even this limited authority was not given to the Administration until 2007, when it was included as a provision of the Food and Drug Administration Authorization Act (FDAAA) and when the escalation of opioid overdoses was already well underway.

It is important to note, as Knight et al. (2017) have done in their examination of pharmacovigilance's clinical-social history, that pharmacovigilance is not just a regulatory tool but also a social technology. And in the opioid epidemic, it has likewise produced rather extreme social effects. In the case of OxyContin, pharmacovigilance and the limits inherent to the FDA's regulatory power over it set the stage for Purdue Pharma to release a wave of relatively unsafe and ineffective opioid products onto the market, where they would migrate from pharmacy shelves into medicine cabinets everywhere at an alarming speed. Pharmacovigilance, moreover, also enabled Purdue to assert their control and begin compiling pharmacovigilance data in ways that would effectively pull the wool over the FDA's eyes and render the company itself the sole arbitrators in OxyContin's postmarket regulation. Before looking specifically at the ways in which Purdue made strategic use of the FDA's ignorance vis-à-vis the pharmacovigilance data it collected and constructed about OxyContin, it is worth noting how and why data gleaned from opioid pharmacovigilance is limited and, as such, paints a partial picture of the reality of painkiller use that fails to reflect the realities of it in real time, a limitation that can and has been used in Purdue and OxyContin's favor.

First, opioid use data is typically compiled at infrequent intervals, which means that by the time a problem has been identified, it would have likely already expanded in both scope and scale. Moreover, the data itself has tended to be limited to emergency room admissions and hospital records, and, namely, those in metropolitan areas, which suggests that data from more remote, rural areas is missing from these technologies. This shortcoming is particularly important for opioid use, since we now know that it is precisely in rural areas on the East Coast and in the Southeast where patterns of opioid use first took shape, and which, until recently, proliferated at much higher rates than they did in urban or metropolitan areas (Van Zee, 2009).

Additionally, most databases collect data on overdoses that are not capable of distinguishing between the different drugs that may have been responsible for them. Databases tend to rely on autopsy reports and chemical analyses, both of which identify the chemical compounds (such as morphine, hydrocodone, or oxycodone) involved an overdose but stop short of identifying the specific medication (or brand) that caused the victim's overdose or death.[5] This is complicated by the fact that, as many studies have shown, it is common practice for people who use drugs (PWUD) to use multiple drugs at the same time (Calcaterra et al., 2013; Gudin et al., 2013; Jones et al., 2012). People who use opioids, for example, have been shown to often combine opioids with alcohol or benzodiazepines—depressants that enhance the relaxing, euphoric effects associated with opioids (Calcaterra et al., 2013; Gudin et al., 2013). When an overdose victim has more than one chemical compound detected in their system, it is nearly impossible to distinguish which substance was responsible for that person's overdose.

In 2002, DEA officer Laura Nagel attempted to present evidence of OxyContin abuse using data compiled from 1,300 death reports: After the data was cleaned and irrelevant cases were excluded, the analysis showed that OxyContin itself was responsible for at least half of all the overdoses. Nagel presented these findings to a group of Purdue representatives and FDA officials, who discounted the data as nonessential since it did not show that OxyContin on its own was dangerous. This is because many of the deaths appeared to involve multiple drugs, including OxyContin. As such, this evidence was not sufficient to reliably conclude that Purdue's drug was the one at fault. *New York Times* reporter Barry Meier spoke to an FDA official who echoed this understanding, asserting that after taking a close look at the available information, "We don't believe there is cause for panic" (Meier, 2002, 2013). And yet, this statement was made in stark contradiction of the fact that by the year 2000, OxyContin was already wreaking havoc on (and causing panic among) a host of communities scattered across the country (DEA, 2001; GAO, 2003; U.S. Government Printing Office, 2002).

Given this, it may come as no surprise that pharmacovigilance plans have been characterized by some—including former FDA Commissioner Scott Gottlieb—as a cop-out, a backdoor route that the FDA uses to allow valuable drugs to get to market, even in cases where concerns about the safety of those products might otherwise result in their rejection (Gottleib, 2007). In a case like OxyContin, where obvious risks exist but the extent to which they will become widespread is unknown, the FDA was able to go ahead with its approval of the drug on the condition that its sponsor begin developing a pharmacovigilance plan to monitor its use. This kind of probation

---

[5] This is particularly relevant in the case of OxyContin because, as we will see in the following chapter, this particular brand (and the way the brand was built) conditioned the widespread uptake of it among consumers.

sentence helped set OxyContin free and, in the aftermath of the first signs of its social effects, helped to safeguard Purdue and the FDA from criticism and enable them to assure the public that a security system was in place to maintain order and to steady what might already have appeared to be a rapidly sinking ship.

This benefited Purdue, which began collecting profits from OxyContin as soon as its drug entered the medical marketplace and subsequently developed and deployed its own pharmacovigilance program that, perhaps to appease the FDA's growing concerns about OxyContin, conveniently worked to produce a picture of opioid use that appeared to be much more contained and much more manageable than it actually was. Moreover, the pharmacovigilance data that was collected by Purdue not only was useful for appeasing the FDA but also provided the company with valuable information about OxyContin prescribers and consumers, which it could then use to target its marketing efforts. This is, in fact, exactly what happened when Purdue agreed to implement its own pharmacovigilance plan, the Researched Abuse, Diversion, and Addiction-Related Surveillance (RADARS) system (GAO, 2003). RADARS not only helped the company strengthen the risk management plan it was encouraged by the FDA to create but also enabled it to collect huge amounts of information about the precise areas (down to three digits of the zip code) where OxyContin use was on the rise. The company also had access to a large stream of data that showed how individual doctors in these areas were prescribing OxyContin. This information, which came from the global health information service, IMS Health, was then used to identify the highest and lowest prescribers of OxyContin and to adjust the company's marketing plan accordingly (Van Zee, 2009).

The same surveillance technology Purdue used to target its promotional efforts also functioned as a tool that the company could use to manipulate the appearance OxyContin abuse patterns in discussions with regulators and law enforcement. In two separate analyses of industry–regulator relations, Nancy Langston (2010) and Alan Brandt (2009) showed how companies have historically attempted to develop new technologies to produce information (whether scientific research or, in the case of opioids, streams of consumer data) that, when combined with public relations efforts, manufacture confusion and doubt among regulators. Purdue Pharma's promotion of the sophistication of its RADARS tool aligns with this historical tendency. The company's surveillance system, which boasted the ability to track OxyContin consumption trends down to three digits of an area's zip code, also succeeded in producing data that appeared more localized than it really was. When the company was called upon in congressional hearings to attest to the abuse potential of its product, it was armed with this data, which it then used to defend the safety of its product and to characterize OxyContin abuse as a very specific, regionalized issue, rather than a nationwide epidemic.

The case of pharmacovigilance, RADARS, and the FDA's mandate that Purdue put together a system designed to produce partial knowledge that limited the FDA's (and the public's) view of the true expanse of the opioid crisis comprise an example of strategic ignorance at its best, where the production of limited knowledge conveys a reality that works in the interests of regulators and industry actors. Through the structural components of regulation, namely, those that delimit (and severely limit) the FDA's powers and responsibilities in the postmarket regulatory context and those that enable the FDA to "allow" drug makers to devise their own pharmacovigilance programs and tailor their own data, it is both structural and strategic. It is structural in part because of the FDA's limited resources and abilities to identify problematic drug consumption trends outside the surveillance systems produced by the manufacturers of those very drugs. But it is also strategic because it hands over the responsibility of pharmacovigilance not to a different FDA branch, nor to another government actor, nor to virtually any actor that would be relatively more neutral with regard to the interests of OxyContin consumption trends, but, rather, gives control over to the drug's manufacturer, the beneficiary of its sales. That manufacturer is then able to take charge of monitoring prescription and consumption trends, using them to further promote its products, and reporting whatever data it sees fit to share with regulators.

Pharmacovigilance is an important regulatory technology that has given shape to the opioid epidemic. It is through this technology that Purdue Pharma was incentivized to produce knowledge about OxyContin that was, at best, biased, and at worst, outrightly manipulative. Using RADARS and the data it produced, Purdue framed OxyContin abuse as a problem tied to specific localities—a claim that went hand in hand with company's assertion that concerns about its dangers, which were circulating among government officials and the public alike, were exaggerated. As such, the company would go on to argue, the need for federal intervention was, at the very least, overstated. Instead, what was really needed to curb opioid abuse was the development of additional drugs to deter OxyContin's misuse in specific settings—drugs that Purdue would, in a short amount of time, offer to create.

And so it was that the FDA, with its knowledge of opioid use severely limited and with carefully skewed data in hand, could only offer weak decisions regarding the future regulation of OxyContin and other opioid-based compounds. One such decision was made in April 2010, amid an ever-rising wave of opioid overdoses. In the face of what appeared to be a looming disaster, the FDA decided to approve the release of a product that it reasoned may be able to provide a solution—a means of mitigating the risks of OxyContin and similar drugs. It resolved to approve a new opioid innovation—a version of "reformulated OxyContin" (manufactured, of course, by Purdue Pharma) that was said to boast special properties designed to deter opioid abuse by "prevent[ing] the opioid medication from being cut, broken, chewed, crushed or dissolved to release more medication" (Markey, 2016). The new "abuse-deterrent" drug, rebranded as OxyContin OP, was soon accompanied by a host of similar products, most of which were also sponsored by Purdue. At that time, the abuse-deterrent formulations (ADFs) were receiving quite a lot of attention, particularly following several studies that evaluated their effectiveness, the results of which seemed promising (Cicero et al., 2012; Severtson et al., 2013; Sessler et al., 2014). In 2013, perhaps as a response to these early studies, the FDA issued a draft guidance document urging pharmaceutical companies to join in the initiative to develop new opioid drugs with abuse-deterrent properties (FDA, 2015). Still today, ADFs are touted by the FDA as a key "public health priority," a distinction that also helps explain why these technologies were also the star of the Administration's 2016 "Opioid Action Plan" (FDA, 2013; FDA Opioid Action Plan, 2016; FDA, 2021). The plan referred to ADFs not only as a public health priority but also as *the* front-running regulatory solution for curbing the opioid epidemic. The Administration even went so far as to make ADFs exempt from certain regulatory measures that were, up to that point, required for the approval of all opioids. No longer would approving these medications require convening an advisory committee. They were, instead, to be pushed onto the market as quickly as possible.

Yet despite the initial reports of the effectiveness of abuse-deterrent opioids, several studies found that figuring out how to bypass the antiabuse mechanisms of the new formulations was much easier than regulators predicted. Dr. Theodore J. Cicero, a researcher who had taken the lead on several of the papers that reported the initial success of Purdue's new drug, explained that in subsequent analyses, he and his team found that "[u]p to one-fourth of people entering drug rehabilitation programs say they have abused the newer version of OxyContin" (NCADD, 2015). And yet, even with these new studies in hand, the FDA insisted (and continues to insist) on their utility as "a step in the right direction" (Reuters, 2010).

Right after OxyContin and similar ADFs began flowing into the market, the Internet lit up with discussions among opioid consumers about how best to use these new drugs. Internet forums dedicated to chronic pain and opioids, specifically, and drug use, more generally, were suddenly filled with discussions and debates regarding the best methods for "defeating" the new abuse-deterrent mechanisms. Online, many people who use opioids recommended dissolving the pills' protective coating in lemon juice or baking them for 10 minutes, while others suggested wetting and reheating them in a microwave until they dissolved and were ready to be injected. According to Bluelight member Guitar-opana-man, who (as indicated in their username) is prescribed the drug Opana (oxymorphone) for the pain accompanying their multiple sclerosis, "The best way to take OPANA ER... is to place the tablet underneath your tongue, lower lip, or next to your jaw line. The medicine will dissolve much quicker, and you should be out of pain in a very short time..." Another person who uses opioids, Jman1982, suggests using the "dremeling" method, which they refer to as "a WONDERFUL way to make the NEW Opana in to [sic] powder." As Jman1982 explains, dremeling involves the use of a Dremel brand drill and requires one to "take a simple dremel with a sandpaper attachment and hold the pill by some needle nose pliers and start sanding the pill..." The entire process, they explained, "takes no more that 60 seconds and there is the powder ready for you to enjoy..." In the same post, Jman1982 also provided instructions for where to find dremeling machinery: "You can get a Dremel for cheap as hell on ebay and they have them at Walmart. You don't need the fancy one, just get the 2 speed one for about $25 or less and get the sanding attachment and your [sic] home free."

Despite the failure of abuse-deterrent opioids, the FDA's reliance on them as solutions to the opioid epidemic belies a regulatory logic that is decidedly noninterventionist. Only rarely has the Administration leveraged its authority to impose limits on the circulation of opioid products, even when faced with mounting evidence of the risks associated with them. In the case of its support of ADFs, the FDA has not acted on its authority to regulate the growth of the opioid market. Instead, it has opted for a laissez-faire regulatory regime, one in which it does not seek to control the growth of the market but, rather, to support its expansion by multiplying the number of products available to opioid consumers, with the hope that some of the new technologies it introduces into this market will deincentivize abuse behaviors. And yet, as we can see from the forum discussions inserted above, and as will hopefully become even clearer in the discussion below, the purported aim of these products to protect people who use opioids by preventing misuse has not been realized.

Today, the failure of ADFs is even more pronounced in light of the FDA's recent actions to stop the circulation of one of its most popular abuse-deterrent opioids. On June 8, 2017, the Administration requested that the drug maker Endo Pharmaceuticals remove its ADF of the opioid Opana (Guitar-opanaman's opioid of choice) from the market. Its decision to do so prompted a wave of headlines, since it marked the first time the FDA requested the removal of an opioid from the market, citing "public health consequences of abuse" and "risks related to abuse" (Christensen, 2017). As the FDA's press release stated, the widespread use of Opana ER posed a public health threat because it was found to be correlated with a massive outbreak of several serious diseases—including HIV, hepatitis C, and thrombotic microangiopathy, a severe blood disorder—among the people who use it (FDA, 2017a).

But how did this happen? The more complex reality of prescription opioid use has much to tell us: In its original formulation, Opana was a popular drug. This was due to different reasons, including the fact that it was strong, long lasting, and therefore relatively cost effective. It was also relatively safe, in part because its high quality meant that PWUD would not have to inject it but could instead simply swallow the pill or take it nasally and receive an enduring effect. Yet when the FDA replaced Opana with a new ADF, Opana ER, in 2012, this all changed. First, the pill's abuse-deterrent updates increased the price of the pill, which meant that many people could no longer afford to swallow or snort it but would have to turn to injecting it in order to get the same effect they were used to. Another way to save money also included consumers splitting their pills with others. The need to share pills combined with the need to inject them resulted in people who use opioids sharing their equipment, including needles (FDA, 2017). But PWUD were not only injecting the pills with dirty needles; they were also having to do so more frequently. This is because unlike the original version, Opana ER transforms into a thick gel when it is dissolved. This property therefore requires opioid consumers to add more solvent to thin the gel out enough to inject it. That consumers need more solvent means that they also need a larger amount of liquid and therefore must fill their needles more than once in order to get a full dose. All these factors—including the repeated use of needles, sharing among Opana consumers, and the sharing of equipment—made Opana ER much less safe in reality than it appeared in theory. Ultimately, the FDA's replacement of Opana with the sophisticated, abuse-deterrent Opana ER, quickly became a recipe for disaster.

And so, in a terrible paradox, the FDA's attempts to steer opioid use in the "right direction" did less to deter abuse than to reconfigure it and push it into new domains, which would prove much more difficult to regulate and police. The unforeseen, albeit catastrophic, effects of Opana ER and other ADFs could perhaps have been avoided had the FDA decided to intervene directly into the opioid market. Yet it did not. This decision, which exemplifies the Administration's noninterventionist approach to opioid regulation, is perhaps best understood as part of a larger context in which pharmaceutical regulation is driven less by safety and security concerns than it is by market interests.

In the case of opioids, the expansion of the market was driven by strategic ignorance and permissive regulatory logics. As we saw in the case of opioid pharmacovigilance, the decision to hand over regulatory authority to pharmaceutical companies anticipated a lack of regulatory action that was based on information gaps produced by pharmaceutical companies, where the knowledge produced about OxyContin abuse was either partial or skewed in ways that would benefit those actors' financial interests. So too, in the case of ADFs, did missing knowledge about the complex reality of painkiller use and the possible effects of these products work strategically on behalf of their manufacturers who were allowed to broadly advertise and sell them to a growing base of consumers.

That opioid manufacturers were allowed to push new opioid products onto the market without advisory committee support and in the face of mounting opioid overdoses is exemplary of a regulatory regime that has been marketized—a regime in which regulatory action avoids intervening directly into the pharmaceutical market and, in doing so, does less to control the actions of drug companies and their products than it does to lend its regulatory power to them. The practical effects of such noninterventionist approach to regulation are, as we have seen, at odds with the long-held principles of the FDA's safety regime. In this context, the question of what drug regulation is and should be is delimited by and enacted through the market and free-market logics.

The FDA's regulatory power is limited by its relationship to market actors and their interests. Its power is also limited by the Administration's historical emphasis on exerting its authority in the phase of regulation before a drug enters the market. Limiting the FDA's regulatory powers to the premarket context produces strategic ignorance by hiding issues that surround the safety of marketed drugs—including the diversity of individual needs and responses related to certain substances. This was certainly the case in the premarket regulation of OxyContin, where the production of limited information originating from the drug's clinical trials helped Purdue's product get to market quickly but crippled the FDA's ability to anticipate its potential risks. In particular, the element of control that structured these trials made it unlikely, if not impossible, to observe "not as prescribed" behaviors and, as a result, to anticipate their proliferation among opioid consumers all over the country.

It is also important to note that the FDA' s regulatory power is also conditioned by the broader political landscape. Researchers have shown how the healthcare battles that raged during the late 1980s and early 1990s were animated by conservative and antigovernment opposition from health insurance companies, pharmaceutical companies, and patient advocacy organizations (e.g., Jasanoff, 1990; Hilts, 2003; Daemmrich, 2004). Throughout these decades, industry and activists formed somewhat "unlikely alliance(s)," whose combined influence helped contribute to the development of formal mechanisms for deregulating the FDA's drug approval process (Epstein, 1996). In the 1980s, the Reagan administration's laissez-faire economic policies coalesced with several drug scandals that left the FDA reeling, trying to recover its reputation and regulatory power. The FDA's struggle to perform its duties in the face of political and economic opposition was amplified in the following decade when its efforts to regulate tobacco companies came under fire from conservatives in Congress. This backlash continued into George H.W. Bush's administration, which implemented revisions to the drug approval process that attempted to reduce regulatory burden on industry (Gladwell, 1992). While these new revisions aimed to speed up the process of getting new drugs through the FDA and into the market, they were not entirely successful and resulted in a series of hearings in which many members of the House of Representatives proposed to dismantle the entire Administration—an idea that obviously did not come to fruition. It did, however, result in the creation of the Food and Drug Administration Modernization Act (FDAMA) of 1997, which provided the FDA with additional resources to fast-track NDAs while slicing in half the average time required for a drug review. Additionally,

FDAMA streamlined the approval process by reducing the number of advisory committees and simplifying many of the regulatory obligations of drug manufacturers. Most importantly, the passing of FDAMA in 1997 solidified ties between the FDA and the pharmaceutical industry, whose interests in getting drugs to market as quickly as possible were formalized as new obligations for the FDA.

The passing of the 1997 Act marks a key event, one that reinforced the bonds that tie the FDA's interests to those of science and industry (Jasanoff, 1990; Epstein, 1996). Likewise, it has also contributed to paving the path that opioid regulation has taken until now, a moment that shows how regulatory practices and processes are not enacted on or over the pharmaceutical market but instead move through it and according to the strategic logics of it. The case of opioid regulation is therefore illustrative of a broader historical tendency in drug regulation, one that is clearly oriented toward marketization. In this way, drug regulation mirrors financial regulation—where Wall Street CEOs frequently become Wall Street regulators. This is exactly what has happened within the FDA, and former commissioner Scott Gottlieb, who oversaw part of the opioid crisis, is no exception. He, like several of the commissioners who came before him, is an individual with deep ties to the pharmaceutical industry and has spent much of his career actively fighting against regulation. It is the case that industry men have occupied the FDA commissioner's office for nearly 40 years: Arthur Hull Hayes, who served as commissioner under President Ronald Reagan from 1981 to 1983, was a well-known industry consultant and had been for years before he finally accepted the commissioner post. Even during his tenure as a top federal regulator, Hayes continued accepting industry honoraria (Hilts, 2003). Following Hayes, the commissioners appointed by President George W. Bush were also industry sympathizers, weak on regulation and with no experience as regulators. Mark McClellan was an economist and free-market enthusiast. Named by the *Boston Globe* as the "FDA's economist-in-chief," McClellan was reportedly loved by the drug industry for his view that the FDA should prioritize new drug development and help the industry in getting new products on the market "faster and more predictably" (Rowland, 2004). McClellan's predecessor, Lester Mills Crawford, resigned from his post after less than a year in office, pleading guilty to a conflict of interest due to his ownership of stocks in several of the pharmaceutical companies he was charged with regulating (Associated Press, 2007). That the FDA has had, and continues to have, a symbiotic relationship with chemical and drug companies is no surprise, at least not while so many of its officers maintain close personal and professional ties to the industry.

It is therefore not a stretch to suggest that today drug regulation is not the province of the federal government alone but also—and perhaps equally—belongs to pharmaceutical companies, for even more than regulators or their advisory committees, it is the companies themselves that decide how their products will be managed at every step of the regulatory process: Pharmaceutical companies not only test and provide evidence of their products' safety and efficacy; they also get to dictate how their drugs will be labeled and how they will be monitored once they enter the market. The entire process of risk management after a drug's initial approval is likewise designed and conducted by pharmaceutical companies. Regulators can only make "suggestions" and provide "guidance" as to how this process should be carried out (Griffin III & Spillane, 2012).

When voting whether to accept the FDA's Risk Evaluation and Mitigation Strategies (REMS) proposal for extended-release opioids, John Jenkins (then director of the Administration's Office of New Drugs), made a point to remind his colleagues of the FDA authority's place relative to industry: ". . . let me remind you that our REMS authority is to regulate the sponsor of the application for the product. So [sic] anything we exercise has to be affected through the sponsor or the manufacturer of the product" (FDA, 2010, pp. 129–130). Jenkins's statement is worth noting: first, because it signals the depth of the ties that run between the FDA and the pharmaceutical industry, and next, because it raises a contradiction, a source of tension between what the FDA takes its regulatory authority to be and the practical effects of this authority: Jenkins's claim that the FDA's authority enables it to directly regulate drug sponsors belies the fact that the Administration also uses its authority to avoid this task. In the case of developing its REMS proposal, the FDA mobilized the authority granted to it by the 2007 FDAAA to delegate several of its own responsibilities to the companies. In doing so, it also made them the implementers of many of the practices that are central to opioid regulation—including the development, monitoring, and management of the various risks associated with their products. Risk evaluation and management strategies are developed and implemented in a way that is relatively free of the FDA's oversight. The passing of the 2007 FDAAA, which gave the FDA authority to delegate a wide variety of regulatory duties to pharmaceutical companies, also laid the groundwork for a regulatory landscape in which tools that once belonged to the FDA—materials for risk communication (labels, medication guides), education, prevention, and promotion—are now conceptualized, developed, and implemented by the very same commercial interests that they propose to oversee. In this marketized world of drug regulation, companies decide how regulation happens, what it entails, and what drugs will be ushered into the market.

Thus, in opioid regulation, regulators' decisions not to intervene directly in the opioid market have been driven by three key factors: First is the strategic ignorance produced about pain and painkiller use in pharmacovigilance and in OxyContin's clinical trials, which limited knowledge about the reality of pain and relief to the controlled context of a trial conducted with opioid-naive patients suffering from acute pain, and not moderate-to-severe chronic pain sufferers. Strategic ignorance has also circulated alongside permissive regulatory logics that define risk only in terms of what is already known about a drug and make it difficult to anticipate future risk scenarios. Taken together, strategic ignorance and a permissive approach to regulation produced a fertile landscape in which opioid manufacturers like Purdue could plant their newest, most profitable products. These two factors also go together with broader historical tendencies toward the marketization of regulation, where private interests and increased consumption override concerns about the potential pitfalls of new opioid combinations.

These three factors can be clearly identified across a variety of congressional hearings and advisory meetings related to opioids, in which FDA committees convened with the objective of determining opioids' abuse potential and developing strategies for monitoring and preventing the spread of risky consumption practices (e.g., DOJ, 2001; FDA, 2010, 2013; U.S. Government Printing Office, 2002; U.S. House Committee on Government Reform, 2005). In these meetings, agenda items that aimed to discuss opioid risk management produced conversations that revolved less around concerns about the possibility of opioid abuse/misuse and more around the idea that any attempt to intervene in the market or manage opioid-related risks would inevitably limit patients' access to pain-relieving drugs. Discussions about how best to balance the individual risks and social benefits of opioids are central to the regulatory discussions about OxyContin and other painkillers and have played an important role in shaping the trajectory of opioid regulation and, as a result, of the opioid epidemic.

The Rise of the Opioid Gray Market

As a result of strategic ignorance, permissive risk logics, and the marketization of opioid regulation, the United States has begun to witness the formation of what I refer to as a "gray market" for opioid drugs. The opioid gray market is useful for showing how micropolitical issues in opioid regulation have paved the way for much larger, structural problems to arise. These issues have helped to create a concrete ecosystem of opioid circulation and consumption that is, at the very least, difficult to control and provides an incredibly hostile environment for the regulation and policing of illegal and semilegal opioid distribution and consumption. In concrete terms, the gray market for opioids has functioned through several key operations: first is the ongoing creation of outpatient pain clinics (or "pill mills"), where opioid products are delivered in bulk and are handed out by hired physicians based on a quick examination and a high price tag.[6] Pill mills are important sites where one can witness the complicated nature of policing opioids since they are technically legal enterprises within which legally questionable opioid prescribing occurs. Another component of the opioid gray market concerns opioid-manufacturing practices that take place on a large scale outside of the United States, where increasingly powerful synthetic opioids are shipped into the U.S. market and distributed and consumed within it. Finally, the gray market is likewise delineated by the ability of synthetic opioid makers to alter the chemical structure of their products so that they can escape DEA scheduling and quickly transform from illegal to legal substances.

On a less tangible and more analytical level, the gray market can also be seen as the cumulative result of regulatory decisions made in a field defined by uncertainty, permissive risk logics, noninterventionist approaches, and strategic ignorance around the question of pain and the drugs that are meant to relieve it. Put another way, we might think of the "gray market" as part of a killer cycle in which ignorance and strategies related to pain and painkillers lead regulators and other key decision makers toward actions that push formerly legal drugs into new, illegal contexts—contexts in which legal medications that have been approved by the FDA circulate but are distributed and/or consumed in illegal ways that, in turn, make them even more difficult to regulate and control.

It is perhaps unsurprising then, that the gray market, insofar as it refers to a regulatory landscape steeped in the chaos of widespread uncertainty and strategic ignorance, has produced unexpected side effects, many of which have prompted disastrous consequences. The production of side effects is

---

[6] Read more about the function of "pill mill" operations in John Temple's book *American Pain* (2016).

not, however, unique to the case of opioids. It is something that critics of U.S. drug policy have long recognized as a problematic effect of drug regulation, which is referred to as the "balloon effect." When you squeeze a balloon, the air inside it may move, but it does not disappear entirely. The same effect occurs when drug regulators attempt to manage risk in one dimension of the pharmaceutical market: the practices they attempt to disrupt are not eliminated but simply move elsewhere or transform in ways that enable them to escape future intervention. This balloon effect is exactly what has taken place in the world of opioid regulation. One only has to think back to the use of abuse-deterrent formulations and the disastrous case of Opana ER to recognize it. But I am also suggesting that the balloon effect in opioid regulation has led not only to new forms of abuse but also to the formation of an entire ecosystem of drug use that escapes regulation and enforcement—a system of exchange that disrupts regulatory logics.

Like the gray market, pain and painkiller use are very "gray" areas. Since we have no way of reliably measuring or defining an individual's experience of pain, there's a world of uncertainty that surrounds it: Is pain a disease or a slippery slope toward addiction? Are opioids useful treatments or dangerous drugs? Are the people being treated for pain patients or potential addicts? Are the doctors who prescribe these treatments caregivers or criminals? Relatedly, the consumption of painkillers likewise refuses to fit into categories that are either black or white, particularly in legal terms—for it is not the case that pain-relieving drugs are either legally received from a doctor and taken "as prescribed" or that they are illegally purchased on the street and "abused." On the contrary, these drugs are used creatively and instrumentally, in ways that may be neither purely medical nor purely recreational. This gray area of opioid use is in part due to the fact that drugs like OxyContin do not always work as they are supposed to. As previously discussed, for many chronic pain patients who were prescribed OxyContin on a 12-hour dosing schedule, the drugs would have proved less effective than their labels suggested, which means that many patients would have had to take more opioids than "as prescribed" or resort to finding more creative ways of consuming their pill in order to get a fuller dose.

According to data from Vowles et al. (2015), roughly 21%–29% of patients prescribed opioids for chronic pain "misuse" them. And 6% of people who use opioids are shown to have transitioned from legal prescriptions to patterns of misuse, and further into the illicit market and heroin use, with a resulting threefold increase in heroin-involved overdose deaths from 2010 to 2014 (Cicero et al., 2014; Compton et al., 2016). In total, 80% of people who use heroin report having misused prescription opioids first (Muhuri et al., 2013).

Yet it is the opioid consumers who legally acquire their drugs but who do not take them "as prescribed" who most disrupt regulatory assumptions, which rely on the distinction between those who "use" and those who "abuse" prescription drugs. Also disruptive to regulatory logics are the contexts of distribution that increasingly characterize the landscape of opioid use. Many of those at risk of overdosing from opioids do not obtain their drugs by handing cash to a drug dealer on a street corner but receive them as gifts from family members or friends (Daubresse et al., 2013). In these contexts, opioid distribution is neither clearly legible as legal nor illegal, but somewhere in between, where it is even more difficult to control.

Taking a look at the patterns of opioid use that have recently taken shape, it looks like the present crisis is increasingly headed in the direction of these gray markets: From 2014 to 2015, opioid deaths increased an additional 15.6%—a swell that is largely attributed not to pharmaceutical opioids such as OxyContin, but to illicitly manufactured synthetic opioids, in particular an illicitly manufactured, synthetic drug called fentanyl (Hedegaard et al., 2017; Rudd et al., 2016).

In a recent article, NPR discussed fentanyl as epitomizing the opioid epidemic's "third wave"—as opposed to its "first wave," characterized by the use of prescription pain medications like OxyContin, and its "second," characterized by shifts in opioid use following the introduction of ADFs and the escalating cost of prescription painkillers (Bebinger, 2019). Now, as drug use shifts yet again into a new, dangerous, and murky terrain, we enter the epidemic's "third wave," marked by the now-infamous fentanyl.

According to recent CDC reports, overdoses involving fentanyl began to rise in 2013. That year, the drug accounted for 1,919 deaths. Just three years later, in 2016, that number skyrocketed to 18,335 deaths. That year, it was the most frequently mentioned drug involved in overdose deaths (Spencer et al., 2019). More recently, in 2018, more than 31,000 deaths involved synthetic opioids, more than the deaths from any other type of opioid, including prescription opioids and heroin. Overdose rates from fentanyl likewise increased across a broad range of populations, including among both males and females, persons 25 and older, non-Hispanic whites, non-Hispanic Blacks, Latinos(as), and metro and nonmetro residents (CDC/NCHS, 2018).

The role that fentanyl now plays in the opioid epidemic is worrisome, not least because it demonstrates that the consumption habits of people who use opioids are shifting toward more potent illicit drugs. Additionally, this movement illustrates a slippage between the worlds of licit and illicit opioid use. As one study showed, at the start of the "third wave" in 2013, nearly 80% of individuals who were then purchasing opioids on the illicit market reported that it was their previous experience using prescription opioids that led them there (Jones, 2013).

Their migration, moreover, has been prompted by recent efforts to manage the market for pharmaceutical opioids—by the increased efforts of U.S. government and law enforcement to police their use. With pharmaceutical opioids increasingly difficult to obtain, more people who use opioids are turning to the illicit market for heroin and fentanyl. For our purposes, the rise of fentanyl is significant as an illustrative example of what the opioid gray market looks like and how it disrupts regulators' attempts to curb opioid use. But it is also worth noting that the movement of drug use toward synthetic forms of existing drugs (which tend to be much stronger than the originals on which they are based) is becoming increasingly popular in general, to the extent that these drugs are reshaping the entire landscape of opioid use. Thus, it may be the case that the concept of the gray market will prove useful not only for those attempting to understand the consequences of opioid regulation, but of drug-related policy and regulation as a whole.

Fentanyl is a synthetically produced opioid medication, created by Paul Janssen in 1960 (Stanley, 1992). In the mid-1990s, it was introduced for the treatment of severe pain—first in the form of a patch, which would allow the chemical to be absorbed directly through the skin. In 1998, the FDA approved another fentanyl medication in the form of a lozenge or "lollipop" that was designed to dissolve slowly in the patient's mouth. Later, the drug was also sold as a sublingual spray. Beginning in 2013, the DEA, along with its law enforcement partners, began noticing an alarming number of overdose incidents and deaths attributed to synthetic opioids—primarily fentanyl and its analogs (DEA, 2017, 3). While pharmaceutical fentanyl in transdermal patches or sublingual tablets was being diverted on a relatively small scale, recent years have seen enormous increases in opioid-related deaths driven by fentanyl products (DEA, 2017, 4). Recent reports indicate that there is not one, but dozens, of fentanyl varieties circulating on the streets (Spencer et al., 2019).

Additionally, it is now becoming more and more common that overdoses occur when fentanyl is combined with other opioids (usually heroin or other synthetic painkillers). These incredibly deadly opioid mixes are sometimes referred to as "gray death," a name that refers to the color of the powder that results from these mixtures—a dull shade of gray, like concrete (DEA, 2017, 6). But fentanyl's most significant claim to fame hits closer to home, in 2016, when the drug was identified as the substance that caused the death of the musician Prince, who overdosed on it in his Paisley Park estate on April 21, 2016 (Eligon & Kovaleski, 2016). Since Prince's death, it has been documented that in certain parts of the country deaths from prescription painkillers and heroin alike have virtually disappeared, having been replaced by fentanyl. In Dayton, Ohio, for instance, 100 drug overdose deaths were recorded in January and February of 2017. Of these, only three were related to heroin. The remaining 97 were caused by fentanyl (Katz, 2017). Additionally, state data from Massachusetts, one of the states hit hardest by the opioid scourge, showed that fentanyl was present in nearly 90% of fatal overdoses in 2018 (Bebinger, 2019).

Fentanyl poses a host of unique problems for regulators, problems that stem from the distinct contexts and practices of production, distribution, and consumption that are associated with it, as well as from the unique characteristics it has: Fentanyl is, by and large, a far more dangerous product than is heroin, morphine, or OxyContin. This is primarily because it is much stronger than any of these other compounds. Its potency is due to the fact that it is a synthetic drug. That is, it does not derive from a plant but is instead developed in a laboratory, where it is designed to mimic the effects of other, natural opioids (such as opium or morphine). Though fentanyl resembles morphine in many respects, it is around 100 times more potent than its chemical cousin and is 50 times more potent than heroin (CDC, 2016). Moreover, the deadliest of the fentanyl analogs—that is, other synthetic opioids that are based on fentanyl—is carfentanil, a chemical whose use was initially intended as a tranquilizer for large mammals such as elephants. Carfentanil is estimated to be 100 times stronger than fentanyl, which makes it 5,000 times stronger than heroin and 10,000 times stronger than morphine (DEA, 2016; Vardanyan & Hruby, 2014). While the lethal dose of carfentanil is unknown, the DEA suggests that just one dose of its much weaker cousin, fentanyl, is contained in an amount smaller than just a few grains of table salt (DEA, 2016).

Even though these drugs can be deadly, it is often the case that fentanyl and its analogs are not readily legible as illicit drugs, which can pose a problem for those agencies charged with stemming their flow. Much of the fentanyl circulating on the illicit market originates as a powder but is shaped by distributers using a pill press so that it ends up resembling a legal, pharmaceutical drug. Oftentimes, these pressed pills even contain the same engravings as they would if you bought them from a pharmacy. There are huge amounts of counterfeit fentanyl pills circulating on the market, even at this very moment. Most of them appear to be other drugs, normally those that are in the highest demand or have recently been removed from the market or replaced with ADFs (such as certain doses of OxyContin, Opana, Vicodin, Percocet, and often even generic, nonbrand opioids like hydrocodone and oxymorphone). These legal-looking illegal drugs are also difficult for regulators to schedule—that is, to place in a category of drugs that are either legal or illegal. While fentanyl remains a legal drug, many of its stronger analogs have been allocated a Schedule I position, which means they are illegal to possess or consume. But because they are produced synthetically, a manufacturer may bypass the limits of drug scheduling by altering one element of fentanyl's molecular structure to create an analog that has yet to be scheduled and is therefore still "legal." For a time then, the drug escapes the grasp of the DEA, while its manufacturers and distributers are safe from legal prosecution and may distribute their unregulated, but technically legal, fentanyl product at will.

It is not only the product itself but also the contexts in which fentanyl is produced and distributed that pose additional challenges for regulators and law enforcement. These contexts are, for the most part, shrouded in grayness since fentanyl may be bought and sold through the nameless channels of the Dark Web. The anonymous circulation of fentanyl products through Internet exchanges have enabled these drugs to arrive at the doorstep of individuals in nearly every region of the country. Most of the time, these drugs are shipped in plain sight and arrive at their destinations in small packages delivered by the U.S. Postal Service (USPS, 2019). Authorities have expressed their frustration in their efforts to put a stop to fentanyl distribution, but the incredible potency of these drugs enables them to move in such small quantities that they are nearly impossible to identify and track (Popper, 2017). The difficulty of tracking fentanyl shipments is also because these drugs tend to be exchanged between individual producers and consumers, unlike other illicit drugs, which tend to be distributed through cartels—larger entities that are easier to track and prosecute. The exchange of fentanyl, in contrast, is more difficult to pinpoint: Instead of a cartel smuggling vast shipments of product across the national boarder and handling huge amounts of cash, picture a tech-savvy individual who sits quietly in their home, buying (or perhaps selling) fentanyl using a small amount of Bitcoin and an old PC. While it would be much easier for the DEA to identify and catch an individual who ships a kilogram of heroin to their house every month, identifying the individual who repeatedly buys or sells small quantities of fentanyl online is far more difficult.

Finally, it bears noting the difficulties that accompany fentanyl's unique context of production—which often takes place in laboratories, most of which are not located in the United States, but in other countries, like China. While the Chinese government has attempted to regulate or, in some cases, ban the production of fentanyl products, it has largely failed to do so. This being the case, there remain many Chinese laboratories that routinely develop and distribute huge amounts of fentanyl in powder form overseas—either to Mexico, Canada, or directly to the United States—in small packages that present a small risk of being detected (particularly if their content has been chemically altered so that neither drug dogs nor other kinds of detection technologies can yet identify it). And when the product reaches its destination—even if it is a very small package—it can easily service a thousand different customers (Ganim, 2017).

* * *

To understand how and why the opioid epidemic looks the way it does, it is essential to examine the regulatory principles and logics that condition the FDA's decision-making practices. Yet even with this knowledge in mind, it is still difficult to comprehend how such a high level of FDA-approved destruction could be reached so quickly and why it has yet to be mitigated. One tool that can help us account for these questions is strategic ignorance—or the systemwide production of limited knowledge that has, at nearly every step of the regulatory process, blinded regulators to the realities of pain and painkiller use. What is clear in this chapter is also the destructive, albeit symbiotic, relationship between institutional logics and industry interests, which is undergirded by strategic ignorance, and which requires further interrogation. If we do not find new ways of dismantling this relationship, it will continue to dominate the institutional landscape of opioids and contribute to the disastrous cycle that feeds the opioid gray market and propels the overdose crisis forward.

Strategic ignorance, permissive regulatory logics, and the marketization of opioid regulation are the principal forces that undergirded the FDA's decision to approve OxyContin based on clinical trials that suppressed critical information about the realities of painkiller consumption and the effects of these drugs in different patient populations. The FDA's sorely limited understanding about how OxyContin could (and eventually would) be used also served as the basis for its recommendation that the drug's manufacturer be the one responsible for creating a pharmacovigilance plan to monitor its use. The plan, which was intended as a way of mitigating unknown risks by detecting early warning signs, ended up obscuring more than it ever showed. For rather than using the data it produced to alert the FDA of the rapidly growing rates of OxyContin overdose, Purdue Pharma chose instead to present its data in a way that would paint a picture of OxyContin abuse as a highly regionalized problem, one that barred the need for nationwide intervention. Finally, it was this strategically produced ignorance that gave the FDA cause to encourage opioid manufacturers to deter potential abuse by developing abuse-deterrent formulations—a decision that ended up reconfiguring the terrain of opioid use and pushing many people who use opioids toward fentanyl and the gray market. It is this new landscape of synthetic opioid use that regulators must now confront—though it presents a host of new problems that, sadly but not surprisingly, the tools they have at their disposal cannot hope to solve.

3
Branding Pain Relief

Though medicine and regulation are crucial starting points for understanding how the problem of pain has conditioned many of the logics that help explain the scope and scale of opioids' uptake in society, the analysis does not stop here. What's also important to understand is how pain and its marriage to opioids became normalized outside of medical and institutional domains and within American culture. For it is also in culture and the American imagination where the opioid epidemic has found fertile ground for growth.

Yet this is easy to forget, in part because much of the discourse surrounding the opioid epidemic has tended to characterize it as a medical phenomenon—as waves of pain patients suddenly transforming into addicts. While this idea is a powerful one that undergirds and animates the anxiety that now surrounds the use of prescription painkillers, it does not reflect the reality of this country's opioid problem. In fact, it is a far cry from the data, which reveals that less than a third of first-time opioid consumers obtained them by being "patients"—that is, with a physician's prescription.[1] In contrast, most people who use opioids initially received the products from friends or family members (49%) (Daubresse et al., 2013). So, if we are to try to account for the prolific presence of opioid use and overdose in this country, we should turn our attention to consider not only how opioids became normalized in medicine, but also how they became integrated into the everyday lives of so many Americans—many of whom were never really patients to begin with. Seemingly overnight, opioids have become household objects, objects that anyone might stumble upon even in the most innocuous of spaces—under the seat of a contractor's pickup, in the office of a school nurse, in Dad's medicine cabinet, Mom's dresser, Grandma's purse.

---

[1] In this context, "abuse" refers to taking a drug in a way other than "as prescribed." In other studies, it has been defined in terms of the following four criteria: "1) use to get high; 2) use in combination with other drugs to get high; 3) use as a substitute for other euphorogenic drugs of abuse; and/or 4) use of the drugs to treat opioid withdrawal" (Cicero et al., 2005, p. 664).

The speed and extent to which the meaning of opioids has shifted—from utterly taboo in the 1970s to a dangerous medicine of last resort in the 1980s and 1990s to an ordinary household object by the mid-2000s—are jarring. And yet, this shift has not occurred by accident. There exists a long history of pain-relieving drugs being incorporated, through branding and other marketing tactics, into everyday life. This was, in the 19th century, the case with morphine and heroin, just as it was with aspirin (Mann & Plummer, 1991). The commercial success of opioids and pain-relieving drugs has been paralleled by their uptake within the home. And it is my contention that the historical tendency of domesticating painkillers has carried into the opioid epidemic, where the normalization of opioids in everyday life has been driven, in large part, by the branding of pain relief, most notoriously through the aggressive marketing of the blockbuster drug, OxyContin.

Building on Chapters 1 and 2, the pages that follow argue that the increasing willingness of physicians to prescribe opioids to their patients, which stemmed in part from mounting awareness of the problem of chronic pain undertreatment, carved out enormous space for a prescription opioid market. In this opening, pharmaceutical companies began to intervene and contribute to the growth of that market. They did so, namely, through the strategic branding of pain relief and pharmaceutical solutions for it.

As many scholars have noted, branding functions as a powerful commercial tool and a cultural process that helps foster relationships between people and products and, in doing so, guides the behavior of potential consumers (Elliott, 2003; Foucault, 1991; Miller, 2008; Rose, 2006; Swenson, 2010). The branding of OxyContin signals a desire on the part of opioid manufacturers to govern the behavior of potential consumers—which they have done by mobilizing the meaning of pain and pain relief as a tool both to sell their products and to attempt to respond to an overwhelming wave of criticism regarding their role in the ongoing crisis of opioid overdoses and deaths.

It should come as no surprise that amid so much death and destruction that has been traced directly to pharmaceutical opioids, drug companies have recently begun to bear the legal and financial brunt of the crisis they are accused of creating. And while this book is not predominantly geared toward a criticism of the pharmaceutical industry alone (for there are—as I've previously argued—many other actors, institutions, and forces at play here), one cannot simply recount a narrative of pain and opioids without casting one's gaze at big pharma. For one of the first times in recent history, major opioid manufacturers are finally paying up for the many indiscretions, manipulations, and crimes with which they have been charged regarding their treatment of painkillers, physicians, and vulnerable consumers and chronic pain sufferers. After a series of lawsuits waged against the most notorious of the painkiller producers, Purdue Pharma, including a 2007 federal case that led to the company dishing out $600 million in fines (Zezima, 2019) and a series of lawsuits that resulted in Purdue paying out another $260 million (Feely & Melin, 2019), the company seems to be finally reaching some kind of breaking point. As a result of these trials, and as part of its plan to try to settle the litigation brought upon it by more than 2,000 different counties, municipalities, and Native American governments, the once small, family-owned business-turned-pharmaceutical-giant has been forced to hang its head and, on September 15, 2019, filed for bankruptcy (Hoffman & Walsh, 2019; Joseph, 2019). And finally, after it did so, the company pleaded guilty on November 24, 2020, to felony counts that included defrauding the federal government and paying off physicians to bolster dispensing of OxyContin. As part of that plea, Purdue agreed to pay $8 billion to the United States (Benner, 2020).

In the widely covered Oklahoma trials, Purdue, Teva, and Johnson & Johnson were all forced to pay for the "greed and avarice" that, in the words of Oklahoma attorney general Mike Hunter, is what motivated them to engage in the manipulative and illegal marketing activities that renders them "responsible for the opioid epidemic in our state" (Hoffman, 2019). The companies, which, respectively, shelled out $270 million, $85 million, and $572 million, were charged with intentionally playing down the dangers of their painkilling products, overselling their benefits, and, in the case of Purdue, using its U.S. Food and Drug Administration (FDA)-recommended pharmacovigilance program and other marketing schemes as deadly techniques that enabled the company to target and direct its distributors to specific prescribers, regions, and patient populations known to be vulnerable to addiction or that contained heavy consumers of OxyContin and similar drugs (e.g., *The State of Colorado v. Purdue Pharma*, pp. 25, 37, 64, 68; *Commonwealth of Massachusetts v. Purdue Pharma*, pp. 10–12, 16–34).

The legal critique of these manufacturers, along with the financial punishments they will now endure (but, if history is to repeat itself, will likely bounce back from)[2] is an important step in meaningfully intervening in the

---

[2] New York's attorney general, Leticia James, accused the Sackler family of plans to shuffle away at least $1 billion of undisclosed transfers from Purdue into the family's own accounts in an attempt bury their money beyond the reach of legal actors (McGreal, 2019).

problem of pain and opioids. However, what is certain, in light of the critiques now nationally waged against the Sackler family (founders and owners of Purdue Pharma), Teva, Johnson & Johnson, and other manufacturers and retailers like Walmart and CVS (Grobe Plante, 2019), is that the nation is finally making a stand to raze their manufacturers—and all those involved—to the ground.

This for many, is great news. One cannot help but feel a sense of justice when one reads the statements by attorneys general from Colorado to Connecticut to Ohio, Illinois, Massachusetts, and Tennessee that hold big pharma accountable for the current crisis. Declarations such as this one by Oklahoma's attorney general, Mike Hunter, assert that "greed caused Oklahoma's devastating opioids epidemic" and [that] *all* these companies will "finally be held accountable for the ... worst man-made public nuisance that our state and this nation has ever seen!" (Hoffman, 2019; Randazzo, 2019, emphasis added). Greed, indeed. One cannot deny it. But, without doing so, I'd like to put forward a question, one that none of these court cases answers: That aside from greed and aside from the "manipulative marketing tactics" and "underplaying dangers while overplaying benefits," what were the actual on-the-ground marketing practices that these companies used to lure in such huge populations of consumers? And what was the rhetorical content of the marketing messages they created that garnered so much appeal for opioid prescribers and consumers? In other words, how did the Sacklers, Purdue, and others manage to pull all this off in the first place?

To find out requires taking a closer look at the branding practices of these infamous manufacturers—namely, Purdue Pharma. By examining the meaning-making practices that underlie the promotion of Purdue's most famous product, OxyContin, we can see how branding not only constructs opioids as an effective solution to the problem of chronic pain but also engages in a broader and more insidious practice of promising its consumers something more (Ingraham & Johnson, 2016; Satterfield, 2017).

By channeling the branding of their drugs by focusing on pain relief and by attaching pain and the promise of relief to particularly powerful American tropes, Purdue Pharma's OxyContin constructed a message that convinced many of its consumers that by adhering to a regimen of OxyContin, they would gain a life of pain. And that would inevitably be a life worth living, one characterized by newfound self-determination, independence, and self-enterprise. In other words, the branding of pain relief, through a variety of skillful meaning-making maneuvers, defined painlessness as an integral and necessary component of achieving the American dream. By normalizing opioids as part and parcel of the American dream, Purdue was likewise able to shield itself from years of waves of negative press, lawsuits, and widespread public outcry regarding the dangers its products posed to society. This rather incredible feat may also help us understand why OxyContin continues to be widely prescribed and consumed today.

Governing Opioid Consumers

Branding is not only a crucial mechanism that sustains the growth of different markets but is also a cultural process that organizes meanings and reflects fundamental social and cultural relations. The meanings that are organized and the knowledge that is produced through branding processes create a context for consumption. This context then frames the ways in which consumers relate to and interpret their relationship not only to products, but also to their own experiences, values, and the world around them (Banet-Weiser, 2012; Hearn, 2010). In other words, one of the key functions of successful branding is the shaping of consumer subjectivity.

Branding is particularly significant in the context of the U.S. health industry, which, unlike the health industries in any other country (except New Zealand), allows pharmaceutical companies to advertise their drugs directly to consumers (Angell, 2004; Dumit, 2012; Elliot, 2003; Miller, 2008; Moynihan & Cassels, 2005; Rose, 2006; Swenson, 2010). Branding processes in this context may be understood as a means of "selling sickness," in which companies enroll themselves in the process of designing and defining new disorders, which are then used to hail a growing population of individuals as sick, or potentially sick (Dumit, 2012; Moynihan & Cassells, 2015). But branding in the drug industry not only works toward the promotion of pharmaceutical products; it also attempts to shape the conduct of potential consumers (Elliott, 2003; Miller, 2008; Rose, 2006; Swenson, 2010).

As each of the above scholars has discussed, branding is crucial to the success of pharmaceutical companies, particularly those attempting to bring new products onto the market. This was certainly the case for Purdue Pharma who, in December 1995, was notified that the FDA had finally approved the release of its newest product, OxyContin. The efforts Purdue poured into marketing its new drug were extensive, and they were rewarded by the near immediate success of its product: Just one year after its release, OxyContin had garnered more than $48 million in profits—a number that, by 2001, had skyrocketed to $1.1 billion. It doing so, it surpassed the sales of Viagra, which had formerly been the most popular drug on the market (Van Zee, 2009; Bourdet, 2012).

The immediate success of OxyContin likely had less to do with the drug's superiority than to the dedication and aggression with which the company promoted its use (Van Zee, 2009). Purdue's marketing campaign for OxyContin was seriously impressive—simultaneously wide-reaching and highly targeted. Additionally, it was comprised of a variety of promotional practices that were aimed not only at physicians and patients but also nurses, pharmacists, and insurers. In 2001 alone, Purdue spent $200 million in the marketing and promotion of OxyContin. This expenditure is not uncommon for large pharmaceutical companies, but it was a serious outlay for Purdue, a family-owned business with a much smaller staff (and budget) than many of its competitors at the time (OxyContin Hearings, 2002).

It is no secret that pharmaceutical companies often mobilize a host of different tactics to sell their products. One of the oldest and most common is the use of sales representatives who are paid to travel to hospitals, clinics, and private practices to promote products directly to physicians in their place of work. Moynihan and Cassells characterize this practice in terms of the efforts pharmaceutical companies' make to buy "doctors for doughnuts" by providing them with incentives (free meals, vacations, etc.) to begin promoting or prescribing a particular drug. In 2000, Purdue hired 671 sales representatives to promote OxyContin using a physician call list that included roughly 75,000 to 94,000 names. The team also drew from an eclectic arsenal of sales tactics, which included the distribution of OxyContin-branded hats, toys, golf balls, coffee mugs, luggage tags, CDs, pedometers, pens, and coupons that provided patients with a free 30-day supply of the drug (U.S. General Accounting Office, 2003).

For years, the branding of pharmaceutical products has likewise relied on the strategic mobilization of medical experts and expertise (Rose, 2006). That is, when marketing a new product, companies commonly employ expert "thought leaders" in medicine and related fields as paid speakers, who are then charged with advertising the company's products to professional audiences in academic/professional conferences and workshops. Purdue also made use of this tactic by promoting its new drug in medical journals and through "continued medical education" workshops, which it paid physicians, nurses, and pharmacists to attend. From 1996 to 2001, Purdue sponsored more than 20,000 pain-related "educational programs" (roughly 10 a day, every day of the week) and 40 national pain management conferences in popular vacation destinations (Temple, 2015; Van Zee, 2009). These "continuing medical education" events are known to have a strong influence on physicians' prescribing behaviors (DeJong et al., 2006). Moreover, doctors in most states are required to attend them in order to keep their licenses (Angell, 2004). It should come as no surprise then that Purdue's events were well attended. Five thousand health professionals showed up to these all-expenses-paid events, many of whom were also recruited to join the company's bureau of paid speakers. In just a few years, the company had acquired a dossier of paid speakers that included more than 2,500 physicians and other medical professionals (Temple, 2015).

Purdue also bolstered its marketing campaign for OxyContin by forming partnerships with professional societies, patient advocacy groups, and other drug companies (Ornstein & Weber, 2012; Wydon, 2015). In 2012, the U.S. Senate Finance Committee launched an investigation into the financial relationships between opioid manufacturers and leading pain advocacy organizations and found that Purdue Pharma had "deep financial ties" to at least 26 such organizations (Wydon, 2015). By acting as a benefactor to professional organizations such as the American Academy of Integrative Pain Management, American Chronic Pain Association, American Pain Society, and the National Fibromyalgia and Chronic Pain Foundation, among others, Purdue attempted to harness the legitimacy of their members' expertise to promote its product within the medical community. Additionally, the company forged an alliance with pharmaceutical giant Abbott, to whom it paid half a billion dollars to help promote OxyContin (Armstrong, 2016). This partnership provided Purdue with access to a huge swath of marketing resources, contacts, and the experience of Abbott's sales force.

Finally, Purdue marketed OxyContin in promotional videos and through the creation of "education" campaigns, both of which the company used to sidestep public criticism of its marketing of a highly addictive drug.[3]

* * *

In 1998, Purdue released a video titled "I Got My Life Back: Patients in Pain Tell Their Story" (see Figure 3.1). Two years later, the company created an

---

[3] By 2004, OxyContin had become the most-abused prescription opioid and a leading drug of abuse in the United States (Cicero et al., 2005).

Figure 3.1. A call to healthcare professionals from Purdue's Team Against Opioid Use website.
Note: From https://web.archive.org/web/20150820214519/http://teamagainstopioidabuse.com/. Accessed 5/19/21.

updated version—"I Got My Life Back, Part 2: A Two-Year Follow Up of Patients in Pain" (Lyons Lavey Nickel Swift, Inc., 1997, 1999). The two films were distributed to 15,000 and 12,000 physicians, respectively, and were made available for consumers to purchase through a website Purdue created, named *Partners Against Pain*. Physicians who received Purdue's videos were also encouraged to show them to their patients, who could watch them in the waiting room before their appointment (GAO, 2003).

In both films, the promise of OxyContin is depicted through the personal stories of five individuals suffering from different types of chronic pain, along with their physician, Dr. Alan Spanos. The film chronicles the personal journeys of these individuals through their struggles to find an effective treatment for their chronic pain. After years of failed surgeries, physical therapies, and frustrating doctor–patient interactions, each person claims to have finally found relief with OxyContin.

In her 2012 analysis of branding in contemporary culture, Sarah Banet-Weiser suggests that we think about brands as "a story told to the consumer" (Banet-Weiser, 2012, p. 4). The story that is told in the branding of OxyContin is one centered on the self—in particular, on the loss of an authentic self and the possibility of its restoration. The self is lost because of pain.[4] It is only

---

[4] Elane Scarry has made a similar argument in her book, *The Body in Pain* (1985), in which she suggests that one of the most powerful rhetorical effects of intervening on the body through pain is the unmaking of subjectivity—or, in other words, of the self.

through the elimination of pain, which OxyContin promises to provide, that individuals may hope to find themselves again.

These videos instigated a whirlwind of controversy when it was discovered that Purdue had neglected to submit them to the FDA for review as it was required to do before distributing them. According to a government report published in 2003, the first film, "I Got My Life Back, Part 1," circulated for three years before the FDA finally reviewed (and subsequently recalled) it. Upon reviewing both films, the Administration found that they utilized information that was, at best, seriously misleading. In fact, both videos featured a series of unsubstantiated claims regarding the benefits of OxyContin and frequently minimized the various risks and negative side effects that are associated with its use (GAO, 2003). That Purdue knowingly circulated misleading information about its drug was, in fact, crucial to the development of the OxyContin brand: Despite the legally and ethically questionable content of its two promotional films, both demonstrate how hard Purdue has worked to brand the treatment of pain using brand-name OxyContin as a moral and ethical obligation for physicians, who are explicitly reprimanded for erroneously attempting to "manage" pain when they "should be treating it ... should be making it better" and should use every tool at their disposal to effectively "shut it down." Luckily for these doctors, the films suggest, OxyContin offers the possibility for atonement. To compensate for not having prescribed OxyContin before now (an oversight that the film likens to medical negligence), all these physicians must do is start prescribing it.

The ethical obligation constructed within these films, which demands that physicians treat all their patients' pain with a particular drug, is illustrative of what Nikolas Rose has referred to as the "pharmacological ethic" (Rose, 2006). While some have argued that the use of prescription drugs (or any other health technology) is driven by the ethical principles of self-enhancement and optimization, Rose says otherwise. He argues that the ethic underlying the use of pharmaceuticals is not about optimization but is, rather, geared toward the achievement of authenticity. That is, the motivation for using these drugs is oftentimes less about achieving an optimal state of happiness/well-being than it is about restoring one's true self—an essence that has been, presumably, obscured by the illness, syndrome, or disease from which one suffers. As Rose states, "In a regime of the self that stresses self-fulfillment and the need for each individual to become the actor at the center of their life, these drugs ... do not promise a new self, but a return to the real self, or a realization of the true self" (2006, p. 100). The theme of returning to the authentic self is a theme that runs throughout the two "I Got My Life Back" videos. But what, exactly, constitutes an authentic self? In Purdue's promotional videos, the self that is envisioned is, first and foremost, an economically productive, enterprising self—a self that is sought and can only be achieved through the long-term use of OxyContin.

Throughout the videos, interviews of the six OxyContin patients are filled with descriptions of how the medication gave them their life back. Yet it is interesting to note that the lives they lost are consistently (and often unilaterally) described in terms of livelihood—of "getting back to work" and restoring their former productivity, activity, and financial independence. For Jonny Sullivan, a civil engineer and construction business owner, chronic pain cost him his business and therefore his financial security. Lauren C. shared a similar story, one where life before OxyContin consisted of many lost workdays and the hours she wasted cooped up at home. For Lauren, OxyContin was a "godsend" that got her out of the house and back to work, where she could become "a productive person again." She is shown sitting at her desk, where she takes a break from typing to turn toward the camera and announce that, "Since I've been on this new pain medication, I have not missed one day of work and my boss really appreciates that!"

The ability of opioids to get patients "back to work" is a key theme in these videos, one that is repeated in each of its six recovery stories, which repeatedly describe Purdue's new painkiller as a "godsend," the key to self-improvement, to becoming as "good as I'm going to be." In another interview, patient Mary C. attributes her career success to OxyContin, saying that "Before I had good treatment, by the end of the day I was hobbling . . . now I work until 6, 7, 7:30 at night oftentimes and I'm fine. I feel as if opioids have given my life back to me." Mary also credits OxyContin with getting her a job promotion, an achievement that she explains has made her "feel whole again." Mary's statement explicates an assumption that underlies the stories of the other five patients: For each, what is understood as an authentic life and self was obscured by the intensity of their pain. It is only with this new medication that they have been able to recover these fundamental truths: OxyContin, which got them back to work, has made them whole again.

The whole self that is envisioned in the two films is not characterized solely in terms of its relation to work, but also in terms of its vitality—it is not only productive, but energetic, enthusiastic, and constantly active. Vitality is a reoccurring theme in the videos, which emphasize time and time again the problematic "myth" that opioids such as OxyContin cause passivity, due to their sedative-like effects. Yes, the films admit, drowsiness and "tired feeling" are officially listed as two of the drug's most common side effects. But what the FDA refers to as "common side effects" are reinterpreted in the films to be "common myths," which are then undermined by the real lives of OxyContin patients, who are documented gardening, shopping, working, swimming, kayaking, and living lives that are "anything but passive." Johnny Sullivan, the construction business owner with lower back pain, describes the drowsiness effect of OxyContin as a "zombie myth." On the contrary, Johnny asserts, with OxyContin he is motivated to pursue all manner of projects: In addition to taking the 50 foxhounds he owns to local competitions, Johnny exclaims that "I ride a horse, I ride a motorcycle, I operate a backhoe, I move heavy equipment on my own . . . I go hunting and fishing, we go to the movies— there's never a drowsy moment around here!" The videos' rejection of passivity also extends to physicians who are reprimanded for their hesitance to prescribe OxyContin to their patients, a practice that is framed as a "passive" approach to treating pain. Spokesman Dr. Alan Spanos demands that physicians change their ways and begin treating pain "energetically" and "vigorously"—presumably with OxyContin, which he declares is the "new standard" of pain management.

Though, as these examples illustrate, Purdue's videos sought to construct a link between OxyContin and images of vitality and life, this has not exactly played out in practice. In 2008, before Johnny Sullivan left to go home from work, he called his wife to complain that his medication was making him feel drowsy. That night, Johnny's car ran off the road and overturned after he fell asleep at the wheel. He was killed instantly.

The tragedy of Johnny's death—which his wife later attributed directly to OxyContin—belies the optimistic vision portrayed in Purdue's branding of it. The branding of OxyContin envisions pain relief as something that, once achieved, automatically ignites the self-activating capacities of its consumers, who are themselves assumed to be endlessly energetic and productive— perfect examples of the enterprising self, which is central to the construction of neoliberal subjects.

By articulating this vision to the problem of chronic pain, branding functions not only as a means for cultivating a market for pain-relieving drugs but also as a critical meaning-making technology—one that establishes a dynamic between the domains of a market and the self, with the ultimate aim of guiding the behavior of potential consumers, in this case through the legitimate desire to live a meaningful life. The films address OxyContin patients (and potential patient-consumers) not only as chronically ill, but also as people with potential—potential that can be easily realized through the consumption of OxyContin. These videos, by homing in on the potential of sick subjects, do the critical rhetorical work of redefining chronic pain sufferers not as symbols of sickness and failure but, rather, as potentially productive, enterprising subjects. In doing so, this rhetorical move artfully reframes these individuals as exemplary subjects of the American dream, subjects who might be summed up using Nikolas Rose's words as those "individual[s] striving for meaning in work, seeking identity in work, whose subjective desires for self-actualization are to be harnessed" (Rose, 1999, p. 244).

Purdue's marketing of OxyContin follows the logic of branding, more generally, in that it articulates the promise of its product within the framework of consumers' desires for self-fulfillment, and—even more broadly—as subjects who, once free of pain, can achieve the American dream. Purdue's videos tap into these impulses by showing their subjects engaging in their favorite activities and pastimes—teaching, training race dogs, returning to work with a smile, and so on—and guide the desires of potential consumers through a series of "from-rags-to-riches" storylines that shape a meaningful connection between pain relief and the realization of self-actualization and achievement of each individual's vision of their ideal future. All of these, taken together, successfully channel meaning making into concrete consumption practices. The crucial definitional work performed in the branding of OxyContin and, in particular, in the narratives told in "I Got My Life Back," shape the meaning of chronic pain, chronic pain sufferers, and pain relief. And if that weren't enough, it is also key to pointing out that this work has entailed a rather more tangible and material transformation in the landscape of pain relief and opioid use—namely, through normalizing and domesticating opioid use in the ordinary spaces and daily routines of its newly defined, potentiated subjects.

For Purdue, providing consumers with the vital energy to become enterprising subjects and subjects of the American dream is also part of an investment strategy. And this strategy promises to serve Purdue Pharma's economic interests in the long term. As previously mentioned, Purdue Pharma released not one, but two, promotional videos: the first, which was released in 1998, and the second, which was released two years later in 2000, titled "I

Got My Life Back Part 2: A Two-Year Follow-Up of Patients in Pain" (Lyons Lavey Nickel Swift, Inc., 1999). The concept of a "two-year follow up" is significant for the way in which it positioned the problem of pain and its treatment using OxyContin within the framework of chronic illness. Put another way, while the message of the first film is that OxyContin provides a near immediate solution for the problems associated with pain, the second video asserts that the quick-fix OxyContin offers is one that must be continuously administered.

In an interesting paradox, the notion that OxyContin provides a "quick-fix" for chronic pain is situated in the films alongside the marketing of its unique ability to control pain symptoms for up to 12 hours. Highlighting the long-acting potential of OxyContin can also be understood as a method for extending its hold over individual patients and physicians, whose adherence to a specific consumption routine has no foreseeable end. Thus, if the first video makes it clear how the branding of pain relief, along with OxyContin itself, might function as a powerful meaning-making mechanism, then the second "follow-up" video illustrates how such meaning making might be extended and maintained over time. Branding OxyContin as the drug to "start with and stay with" reflects what Joe Dumit (2012) has referred to as "dependent normalcy," the idea that if an individual is to continue feeling well, enterprising, and potentiated (their "new normal," as it were), then they must also continue consuming pharmaceutical drugs. In line with Dumit, Purdue's promotional videos suggest that, with the continued use of OxyContin, the patient in pain will be able to remain "dependently normal" (Dumit, 2012, p. 171).

The steps required to remain "dependently normal" are also steps that resonate with Colleen Derkatch's elegantly articulated idea about ongoing restoration and optimization. That is, the restoration of one's authentic, pain-free self takes more work than the one-time administration of a pill. On the contrary, it requires the continuous labor of performing a patient role. Since OxyContin is a Schedule II narcotic, one cannot simply go to the pharmacy to refill it at will. To get more pills, a patient must first make an appointment to see a physician and obtain a handwritten prescription, which they must then take to a pharmacy to be filled. Thus, what is required when one decides to "start with and stay with" OxyContin is an ongoing relationship with self-medication. It includes, for the physician, the labor of reassessment and record keeping; for the patient, it is adhering to a schedule of regular appointments, follow-up examinations, trips to the pharmacy, and more, if one is to continue consuming the drugs that they need to maintain the new normal of their life with OxyContin. Moreover, as Derkatch suggests, restoration is often not enough. For one can always "feel better." In fact, most of us want to feel "better than well." This yearning—which, in the context of chronic pain, might look something like attaining the ideal state of complete painlessness (something that rarely, if ever, is the case for any of us)—points to the idea of optimization, a rhetoric and a practice that, when realized, leads to an endless cycle of pharmaceutical or nonpharmaceutical modification. To be truly restored, one must engage in the never-ending work of subjectification—of always modifying oneself through various interventions that promise to make us better than well.

Such is the irony of the pharmacological ethic, in which continually performing a patient role is also the key to restoring one's authenticity. OxyContin patient Dorothy M. touches on this paradox rather directly: By time of the second video's creation, Dorothy is consuming 1,200 milligrams of OxyContin every day (a huge amount in relation to the other patients, who consume between 40 and 80 milligrams a day). She explains that she started taking the drug because "It was important for me to get out of the patient role." And yet, if Dorothy stays with OxyContin, a drug that she describes as enabling her to "find life again," she likely never will.

Another way in which the branding of pain relief has been carried out through attempts to guide consumers' beliefs and behaviors is through the development of "educational" and "awareness raising" campaigns. Marcia Angell (2004) has written that pharmaceutical companies often mobilize "education" campaigns around a particular diagnosis (rather than a drug) to achieve two purposes: First, they do so in order to indirectly market a "cure." Second, companies also promote diagnoses as a means evading legal constraints that restrict the ways that companies advertise their drugs to consumers and limit the uses for which they can be explicitly promoted. Though I do not dispute these arguments, I also want to suggest that such campaigns, which are also prevalent across the beauty and self-help industries, do more than enable a company to circumvent restrictions on its marketing practices. Rather, branding under the guise of education provides companies with an opportunity to rhetorically position themselves as socially responsible and trustworthy corporate actors while simultaneously securing the long-term stability of their market by constructing a base of educated and responsible patient-consumers.

The process of repositioning one's company as trustworthy can be particularly useful as a rhetorical "fixer" strategy, especially when that company has fallen under intense public scrutiny. This was the case for Purdue Pharma in the years following its videos and the release of its heaviest and widest-hitting OxyContin advertising campaigns. By 2000, the company had already begun to witness an outpouring of negative press regarding the addictive properties of OxyContin (e.g., Meier, 2001; Ordway, 2000; White, 2001). In lieu of these reports, Purdue's creation of the education/advocacy websites *Partners Against Pain* (1997) and *Team Against Opioid Abuse* (2015) worked strategically to legitimate the marketing of what was fast becoming an infamous brand of narcotic painkiller.[5] While both websites are technically "unbranded" (i.e., they do not explicitly mention or promote any of Purdue's products by name), they nonetheless function as promotional tools.

On its home page, *Partners Against Pain* defines itself as "an alliance of patients, caregivers, and healthcare providers working together to alleviate unnecessary suffering by leading efforts to advance standards of pain care through education and advocacy" (Home Page, 2004, April 2). One of the ways in which *Partners Against Pain* grooms potential consumers is through the use of assessment tools, which are available for visitors to download directly from the site. These tools include checklists and lists of questions used to guide patients through the process of obtaining medication from their doctors ("Checklist: 11 Ways to Ensure Proper Pain Management") as well as the "Find a Doctor" tool—a database visitors can search to find physicians who are trained in providing pain management and whose offices are located close to the individual's zip code (*Checklist*, 2004, June 10; *Find a Doctor*, 2004, June 10). As of July 2002, over 33,000 physicians were included in this database (GAO, 2003, p. 24). Tools such as these go further than education: They also manage communication and information in a way that is normative and directional and, in doing so, guide visitors' behavior toward actions that benefit the company. They accomplish this by activating visitors' own capacities in ways that attempt to train them how to participate in the market for pain relief. Tools such as the "Find a Doctor" database incite visitors to mobilize their lay expertise in ways that guide them toward the best sources from which they can obtain a chronic pain diagnosis and opioid-based treatment plan. In

---

[5] Though both websites have since been shut down, a glimpse into the past through the Internet Archive's "Wayback Machine" provides us with images of the two domains as they appeared in early 2004 and 2015, respectively.

doing so, they also reinforce a "do-it-yourself" (DIY) ethic of health in which individuals—rather than governments or companies—are called upon to take charge of their own vitality and well-being. According to this DIY ethic (which will be elaborated on in the next chapter), it is the individual and them alone who is empowered to seek out treatment and who is encouraged to do so through their own initiative.

By 2003, Purdue's *Partners Against Pain*, along with its other unbranded websites, were drawing criticism: The FDA expressed concern about the accuracy of and intentions behind much of the information published on the site, which "appeared to suggest unapproved uses of OxyContin for postoperative pain" that were inconsistent with the drug's labeling and "lacked risk information about the drug" (GAO, 2003). Four years later, the company pled guilty to a felony charge of "illegally misbranding OxyContin in an effort to mislead and defraud physicians and consumers" (Statement of U.S. Attorney John Brownlee, 2007, p. 1). Consequently, Purdue was forced to pay a $600 million in a settlement—one of the largest ever paid by a drug company (Meier, 2007).

But then, in 2015, in the midst of an investigation launched by the New York attorney general's office into the company's business practices, Purdue decided to unveil yet another unbranded website, *Team Against Opioid Abuse*, which it characterized in a press release as "a new website designed to help healthcare professionals and laypeople alike learn about different abuse-deterrent technologies and how they can help in the reduction of misuse and abuse of opioids" (Purdue Press Release, 2015). As in the case of *Partners Against Pain*, Purdue's new site was not clearly identified as being associated with the company, except for a small copyright logo at the bottom of its home page.

In a way, *Team Against Opioid Abuse* can be understood as a rhetorical strategy to respond to the mounting concerns and critiques regarding Purdue's business practices and the dangers associated with its products. By the time the site went live, the U.S. Centers for Disease Control and Prevention (CDC) had already declared the existence of a nationwide "epidemic" of prescription opioid abuse (CDC, 2011). And as the maker of the best-selling opioid OxyContin, Purdue was implicated as the "small but ruthlessly enterprising manufacturer" responsible for causing the epidemic (Mariani, 2015).

As its name indicates, the *Team Against Opioid Abuse* website is rhetorically constructed around the theme of abuse prevention—a decision that works strategically to shield the company from the negative press and attention that was being directed toward its products. And yet, this site also functions as a marketing platform for those same drugs, which it frames as a solution to the problem of opioid addiction and abuse. This contradiction—of intending to prevent the abuse of opioids while simultaneously promoting the use of those same drugs—is reflective of similar contradictions within the broader economy of opioid use, particularly as it relates to the emergence of a new consumer demographic: At the time of the site's release, the FDA had just approved the use of OxyContin for pediatric (namely, teenage) populations (ages 11–17). This was a controversial and contradictory decision, considering that another government agency, the Substance Abuse and Mental Health Services Administration (SAMSHA), had recently released a report showing a surge in prescription opioid abuse among this same population, whose rates of addiction nearly doubled between 1994 and 2007 (SAMHSA, 2014; American Society of Addiction Medicine, 2016).

Uncoincidentally, the website's imagery constructs an antiabuse discourse that appears to target that demographic, specifically male teenagers. As its name indicates, *Team Against Opioid Abuse* plays on a sports-based metaphor, one that positions site visitors as part of a "team" competing to defeat opioid abuse. The imagery on site's home page bears a more specific reference to football, a sport that is central to the culture of many high schools and universities across the United States. Centered on the home page is an image of a "team" of healthcare professionals: A doctor and several other healthcare providers are shown standing side by side in a line-up formation, each sporting eye black and a competitive stance—with stares fixed directly on the viewer and arms crossed in front of their chests (see Figure 3.1) (Home Page, 2015, October 11).

Marketing abuse-deterrent opioids using a football theme seems, at first glance, a curious choice, yet perhaps not when one considers that around the same time the site went live, the media landscape was littered with coverage of painkiller abuse in the NFL. The website's design seems to reflect a keen awareness of the fact that opioids had become, according to various media reports, a "football problem" (Easterbrook, 2014; Silverman, 2014). Without claiming to know what the exact intentions behind this design decision were, it suffices to say that one of its effects was to reinforce a link between football and opioids that was, at the time of the site's release, already circulating in the public imaginary. But this association, which was depicted in much of the media's coverage of it in more spectacular terms, is reframed on the site as familiar and, above all else, manageable.

Manageability, in this image, is also linked to its gendered emphasis. Specifically, it recalls and reinforces a gendered discourse related to men and the relationship they are expected to have with pain. Pain, as the gendered narrative goes, is something with which men must not only be familiar but also toughen up to ignore, deal with, or, in some other way, dominate. As various analyses of masculinity and embodiment (e.g., Bordo, 1993) suggest, embodiment for men involves a relationship not of reflection or self-care, but of control. Men are not supposed to let their embodiment impede them but, rather, must exert absolute control over their bodies (a powerful discourse that, as we will see later, is also at work in the rhetoric of pain-related self-help). For men, to avoid the processes of self-reflection and ongoing self-care, the solution to pain must be quick and absolute—a search for immediate relief and repossession of control and mastery over their bodies. These gendered discourses are not lost in Purdue's "Team Against Opioid Abuse" advertisement, which, as previously mentioned, in addition to being released during a media frenzy around pain and opioids as a football (aka male) problem, is visually structured for an audience that is compelled to address their pain aggressively, competitively, without self-reflection, and without the attention to their own self-care, which would undoubtedly make them a bad team member. They must stand up and join the fight head-on. This gendered rhetoric and imagery, though striking, is not speculative; at the time of this site's release in 2015, men were opioids' biggest consumer demographic, with the number of opioid deaths among men nearly doubling those among women (Kaiser Family Foundation, 2018).

In the above image, situated on either side of the physician-led football team are links that lead visitors to various kinds of technical, specialized content (regulatory documents, prescribing guides, CDC data sets, etc.)—a suggestion that the manageability of opioid abuse depends on the cultivation of the visitor's own expertise. The site calls on its visitors—physicians, pharmacists, and patients alike—to become experts in the fight against opioid abuse and diversion.

This call to action also signals the site's second purpose—the promotion of Purdue's newest products, opioid drugs with newly developed "abuse-deterrent properties." While the company does not explicitly mention any of its brand-name products, it is nonetheless promoting them: For when the website was created, there were only four abuse-deterrent opioids on the market, all but one of which were manufactured by Purdue (FDA Facts: Abuse-Deterrent Opioid Medications, n.d.). However, the cultivation of this new market did require Purdue to compete—this time not against the specter of addiction, but against the numerous pharmaceutical companies that manufacture opioid products without these new properties. On the website, a page with the header "What's Your Role?" addresses different types of consumers and the "part you can play in the effort to reduce opioid abuse." In each case, physicians, pharmacists, patients, and third-party payers are advised to change the ways in which they practice medicine and participate in the opioid market. They are encouraged to do this by refusing to use any opioid other than those sporting the newest abuse-deterrent technologies.

On the company's website, Purdue specifically refers to its development of abuse-deterrent drugs as constituting a "corporate responsibility" initiative— one aimed at curtailing opioid abuse and diversion. In doing so, the company grafts its marketing practices onto existing social problems: The epidemic of opioid addiction and overdose, which the company insists is the reason for website's creation, provides it with a rationale for the marketing of its new products.[6]

Purdue refers to its development of abuse-deterrent drugs as constituting a "corporate responsibility" initiative—one aimed at curtailing opioid abuse and diversion (Opioids & Corporate Responsibility, n.d.). One might consider the company's claims of advocacy and responsibility to be a form of what Mukherjee and Banet-Weiser (2012) have termed "commodity activism"—a mode of activism that involves "grafting philanthropy and social action onto merchandising practices, market incentives, and corporate profits" (p. 1). Though for Mukherjee and Banet-Weiser, commodity activism is used to characterize activist practices that are typically taken up by consumers, these practices can also be utilized by companies and corporations. For Purdue, commodity activism is another strategy that enables the company to reframe itself and its business practices within a moral framework. In doing so, it also works to shape the conduct of potential consumers (Elliot, 2003; Miller, 2008; Rose, 2006; Swenson, 2010).

As with its older website, *Partners Against Pain*, Purdue's newer advocacy platform also emphasizes the importance of "education" and, to

---

[6] It is also worth noting that Purdue has recently been granted a patent for a new drug to wean addicts from opioids, representing yet another move to reframe the company's practices under the banner of corporate social responsibility—while still driving up company profits (Bever, 2018).

that end, provides its visitors with a variety of instructive materials. On the *Team Against Opioid Abuse* site, it is healthcare professionals who are called upon not only to see themselves as responsible actors, but also to utilize the tools provided on the site to "train" themselves to perform their jobs differently—in this case, by using new abuse-deterrent products. For instance, one of the hyperlinks embedded on the site's home page leads to a review of the FDA's "2014 Guidance for Industry" mandates, which explain the different methods one can use to identify which opioid formulation boasts the latest FDA-approved, abuse-deterrent properties and which does not (*FDA Guidance for Industry*, 2015, December 12). Healthcare professionals are guided through the FDA document to a link located at the bottom of the page. The link then leads them to a selection of videos, press releases, and other documents that tout the potential of abuse-deterrent formulations to reduce the prevalence of prescription drug abuse.

The tools that *Team Against Opioid Abuse* provides to train its visitors form what Foucault (2009) might have characterized as "governance-at-a-distance," where the guidance of individual behavior (in this case, of consumers, patients, and physicians) works indirectly through market mechanisms. In this case, branding establishes a relay between the aims of authorities—like the government and pharmaceutical companies, which, respectively, aim to prevent drug abuse/diversion and to sell their products and protect their reputations—and the aspirations of autonomous citizens (Rose, 2006). But what, exactly, are visitors "training" to become? Better citizens? Perhaps. Better physicians? Some. Better consumers? All—for this particular kind of training does not end with the cultivation of a well-educated public. Instead, it ends with a public that is also seen as a potential market, a population that has been groomed to exercise choice in a specific way, by becoming responsible opioid consumers.

Governing Clinical Practice

The branding of pain relief has not been limited to the creation of Purdue's videos and websites. On the contrary—much of the information and strategies within these spaces have been extended into institutional settings, where they have become embedded in the logics and practices of pain management in hospitals and clinics.

In 2001 and 2002, Purdue Pharma funded a series of nine programs across the country to extend its "education" efforts into hospital settings. The programs were centered on teaching hospitals and other healthcare organizations how to comply with the new "Pain Standards" that had recently been implemented by the Joint Commission for the Accreditation of Health Care Organizations (JCAHO; GAO, 2003). JCAHO is perhaps the most widely recognized accreditation body in the country. It is responsible for the accreditation and certification of tens of thousands of healthcare organizations throughout the United States. Accreditation is a key component of a healthcare system where the administration of quality assurance and standard-setting programs is conducted by entities that are, for the most part, independent of government control. Moreover, hospitals and clinics are incentivized to pursue state licensure through JCAHO's accreditation process, as doing so enables them to receive federal reimbursements for Medicare and Medicaid expenses.

Through its partnership with JCAHO, Purdue Pharma effectively transplanted the ethic of the enterprising, well-trained self that was central to its branding of pain relief in the "I Got My Life Back" videos and on its websites into the U.S. hospital system. Remarkably, Purdue was one of just two companies that were given the opportunity to sponsor JCAHO's pain management training programs. Yet of these two companies, Purdue was the only one allowed to distribute its own branding materials among JCAHO-accredited health organizations and on the organization's website (GAO, 2003).

While a partnership between a pharmaceutical company and an independent organization such as JCAHO may not, at first glance, seem out of the ordinary, it is significant. Purdue's partnership with JCAHO facilitated the company's access to tens of thousands of hospitals and healthcare organizations across the country. In doing so, it extended the reach of the company's marketing campaign into these domains (GAO, 2003). This practice of dispersing branding materials into new contexts is emblematic of what Joe Dumit (2012) has referred to as "strategic ubiquity," a tactic in which companies attempt to create a universe of syndicated and sponsored content by forming alliances with advocacy groups and developing partnerships with other influential third parties. For potential consumers navigating this landscape, every piece of information they read or hear about inevitably directs them toward specific actions that will serve the benefit of that company. Purdue's alliance with JCAHO goes a significant way toward crafting a universe of branded pain relief. As of 2001, the patients sitting in the waiting room of any one of the thousands of organizations that adhered to JCAHO's pain management standards would have likely found themselves reading brochures and watching videos that were created by Purdue with the intention of selling its latest pain-relieving products.

Also in 2001, JCAHO began requiring hospitals seeking accreditation to adhere to the requirements of its newly released pain management standards. These standards, which Purdue helped develop, mandated that accredited organizations use educational tools (namely, those developed by Purdue) to educate health practitioners, along with patients and their families, about the importance of pain management (JCAHO, 2001). Additionally, the new standards included a requirement that physicians begin administering patient satisfaction surveys to every one of their patients before discharging them—to begin polling these patients about the degree to which their pain had been adequately acknowledged, assessed, and treated. The results of these surveys were then used to dictate whether a hospital would be eligible to receive Medicare and Medicaid reimbursements. A 2013 survey of 182 healthcare organizations in the United States found that 66% also relied on patient satisfaction surveys to rate the performance of their staff (Falkenberg, 2013). For physicians, the scores they received from these surveys became key variables for determining whether they would receive extra compensation or, potentially, lose their jobs. As Forbes reported in the same year, nearly two-thirds of physicians had annual incentive plans tied to the scores they received from patient surveys (Falkenberg, 2013).

The dominance of patient satisfaction as a metric for care is closely connected to the development of arguments about pain undertreatment, which has served as a basis for implementing patient satisfaction surveys in hospitals across the country. Yet this approach to undertreatment has produced negative consequences for both medical professionals and people who use opioids. For medical professionals, the use of patient satisfaction as a surrogate for physician or hospital quality presents an unsavory choice of either prescribing opioids or facing the consequences of low satisfaction scores from disappointed patients who may have wanted them. The pressure patient satisfaction places on medical professionals to prescribe or be punished has been posited by some public health experts as having led to the overprescribing of opioids and to growing rates of opioid abuse and overdose (e.g., Adams et al., 2016; Carrico et al., 2018). While patient satisfaction standards are surely not the only factor underlying the opioid epidemic, it is also the case that in the two years following the uptake of these new pain standards, the United States saw steep increases in both the number of opioid prescriptions written by physicians and the average rate of patients' consumption of these drugs (Frasco et al., 2005). This surge in opioid prescriptions, though it may be consistent with the process of normalization that these drugs underwent in medicine in the 1990s, occurred even while public awareness grew regarding the risks that accompany long-term opioid use.

However, to say that new clinical care metrics like patient satisfaction have had negative social consequences is not to say that pain undertreatment is not a problem. On the contrary—it is certainly a problem, one that has become particularly visible ever since pain organizations and key health institutions began tightening opioid prescribing guidelines during the epidemic. Undertreatment is especially common among people of color, people with a history of substance abuse, and other socially marginalized groups who use opioids. Members of these groups have often, throughout history, been refused opioid treatments for pain at much higher rates than white, privileged patients, as several studies have shown (Netherland & Hansen, 2017; Friedman et al., 2019; Anderson et al., 2009; Green et al., 2003; Hoffman et al., 2016).

Yet the ways in which JACHO and other health institutions have gone about addressing the issue of pain undertreatment have proven to be more than a little problematic. The widespread use of patient satisfaction surveys, for example, is problematic because it transforms pain management into a practice that is defined by market logics, which leads to the dominance of market-based solutions for the problem of pain, effectively erasing pain's social dimension. Or rather, instead of seeing pain and pain relief as manifestations of the complex biological, psychological, and social conditions that impact patients' lived experience of them, pain management driven by patient satisfaction envisions pain relief as a commodity, which is experienced in the same way by everyone and which anyone can—and should—be able to obtain in a privatized health system.

Additionally, the concept of patient satisfaction mischaracterizes patients as customers and succumbs to the notion that the customer is always right, which undermines medical authority and resituates the focus of healthcare from making patients well to making them happy. That pain treatment would be dictated by market logics rather than medical/scientific evidence and the social determinants of health renders both patients and their physicians as market actors. They become players in a system that is not concerned with the complexity of human suffering, so long as customers get what they want, and so long as key market actors continue to prescribe and purchase medical products. And in an even larger sense, the use of patient satisfaction as a benchmark for deciding whether a hospital will receive reimbursement has also been shown to increase health disparities, since safety net hospitals, which serve under-or uninsured communities, are more likely to score poorly on patient satisfaction measures (Chatterjee et al., 2012). They are therefore more likely to receive less money, which means the populations they serve will be less likely to receive quality care in the future. Yet this is the market-driven logic of patient satisfaction in pain management, which, as Goldberg and McGee (2011) have shown, is part and parcel of a for-profit healthcare environment.

In addition to patient satisfaction, the growing popularity of the concept of pseudoaddiction—which was developed in the late 1980s and popularized in pain management throughout the 1990s—offers yet another example of the ways in which market logics have driven (and continue to drive) pain management and the social consequences that this has produced. Pseudoaddiction, a concept presented by David Haddox and David Weissman in a 1989 paper in the journal *Pain*, "describes the iatrogenic syndrome of abnormal behavior developing as a direct consequence of inadequate pain management" (Weissman & Haddox, 1989, p. 363). Pseudoaddiction occurs when a physician or another member of a pain patient's healthcare team mistake certain medication-seeking behaviors as signs of opioid addiction—when they are simply signs that the patient's treatment plan is not working, that their pain is not being properly controlled, and, in turn, that they need more and stronger opioid medication. According to Weissman and Haddox, the behavioral indicators of opioid addiction (asking for early refills, lying about consumption habits, etc.), which result in the false diagnosis that is pseudoaddiction, are not seen as a result of true physical or psychological dependence but as a result of either inadequate pain management or a doctor–patient relationship lacking in trust. In response to these factors, patients may "develop feelings of anger and isolation." And as the authors suggest, these factors will in turn lead to "acting-out behavior" and to frustration on the part of the healthcare team, who "will seek to avoid contact with the patient as a means of reducing the source of conflict." This call-and-response process then becomes a cycle in which the reactions of pain patients and healthcare teams "continually interact until a crisis based in mistrust ensues" (p. 364). To avoid the phenomenon of pseudoaddiction and all the conflict it entails, physicians must work to establish trust with their patients while also being sure that they are consistently providing "appropriate and timely analgesics to control the patient's level of pain" (p. 363).

Though this concept has since been widely disputed in pain management, it is still listed as an "up-to-date" concept under the umbrella of opioid addiction as defined by the Federation of State Medical Boards (Chabal et al., 1998; Greene & Chambers, 2015; Dowell et al., 2013; Vijayaraghavan et al., 2013). In recent years, however, the concept of pseudoaddiction has come under fire for the role it is seen as having played in the upsurges of long-term prescription opioid use that many consider to be the driving force of the opioid epidemic, at least initially. In media accounts of the epidemic, pseudoaddiction is often pointed to as a convenient and disingenuous answer invented by big pharma sympathizers for the many physicians who saw in their pain patients the symptoms of growing tolerance and, in some cases, opioid withdrawal (e.g., Deprez & Barrett, 2017; Kessler, 2017; Radden Keefe, 2017). A concept like pseudoaddiction—especially when presented in the pages of a respected medical journal—also works to legitimate long-term opioid prescribing and encourage the continued sale of opioid products, particularly in cases where it is unclear whether the pain patient's relationship to that product has soured.

Moreover, it is not insignificant that one of the authors of the paper that introduced the pseudoaddiction concept, Dr. David Haddox, signed on as a paid speaker for OxyContin-maker Purdue Pharma shortly after the article's publication. Haddox has since been promoted to the company's vice president of health policy and has been paid to travel all over the country and give talks that simultaneously spread the word about pseudoaddiction and promote Purdue's latest opioid products. Moreover, pseudoaddiction has also made frequent appearances in the marketing materials Purdue released in its promotion of OxyContin, including in educational pamphlets designed to provide physicians and patients with reliable information about the safety of opioids and in the "I Got My Life Back" videos. Thus, while it would be difficult to say, unconditionally, that the understanding of addiction espoused in the concept of pseudoaddiction led to the opioid epidemic, it is certainly the case that such an understanding complemented the interests of the opioid industry, just as it complemented the interests of physicians (most of whom were well intentioned and simply wanted to help their patients find relief) and pain patients (nearly all or all of whom want to be pain free).

* * *

In an age when pain is considered to be the "fifth vital sign," we are all confronted with the obligation to consider the problem of pain—and the possibility of finding relief. Whether we are the patients being asked in a checkup to assign a number to our pain or the nurses who are required to do the asking (tens or maybe even a hundred times a day), the question of pain materializes and extends to all of us in an open invitation. To participate in the question of pain today is to participate in the possibility of finding a solution for it—and the solution that has been for some time now the most accessible and most readily available is opioids.

The problem of pain and its relief and the ways in which it was so effectively branded in the United States have resulted in millions of Americans inviting opioids into their lives. But, as we know, what was initially an invitation has since mutated into a catastrophic invasion. Yet, in less spectacular terms, it is precisely the everydayness of opioids that has enabled their migration into medicine cabinets across the country. The processes through which opioids—as a first-line solution for the problem of pain—have been constructed both as familiar objects and as objects articulated to familiar narratives of happiness, success, and the American dream help us explain the formation of today's opioid "zeitgeist." I use the term "zeitgeist," specifically, to indicate the following key points.

First is my argument, which was asserted at the beginning of this chapter, that instead of seeing the opioid epidemic as the consequence of millions of patients suddenly morphing into addicts, we should instead examine it in terms of the ways in which opioids have become normalized as both ordinary household materials and extraordinary immaterial signifiers—of authenticity, actualization, and inner truth. Relatedly, opioids are a zeitgeist because they cannot be understood only in terms of their materiality but must be considered in terms of their discursivity, their meaning-making potential that has been channeled in ways to set the stage for devastating material consequences.

To that end, it is also important to take a slight turn away from big pharma and the traditional healthcare industry to better understand how the cultural logics, rhetoric, and meaning making around pain take shape—how pain is framed and how it functions. Such an analysis, which attends to the ways in which American selfhood is defined and harnessed in the branding of pain, may also benefit from broadening its view beyond the pharmaceutical industry and looking instead to its antithesis—the nondrug self-help industry.

By looking in a different domain, and no less, one that explicitly pits itself against big pharma and its drug-based logics, we find that the rhetoric that makes pain relief meaningful as integral to the ideal American self—and which frames this idea in ways that have led to a nationwide search for pain relief—is also at work in the places where we may not have expected to find it.

4
Self-Help and the Rise of the Pain Patient-Expert

By and large, companies operating within the self-help industry have contributed to the branding of pain relief through marketing an assortment of nonpharmaceutical products, which have been positioned as inherently valuable for the ways in which they offer consumers "drug-free" methods for managing pain. Yet while this industry's products are positioned in opposition to the pharmaceutical industry, the themes that undergird the branding of pain relief in this context closely resemble those utilized by Purdue in its campaign to market OxyContin: by calling upon individuals to become active participants in their own pain management by taking control of their symptoms, self-help and wellness discourses have also contributed to the domestication of pain relief as both an everyday practice and a necessary requirement for consumers' self-actualization. Put another way, in a move nearly identical to those made by their competitors within the pharmaceutical industry, companies devoted to drug-free "wellness" and "self-help" have attempted to frame the search for pain relief as a journey of self-restoration—of getting one's life back.

Both industries, moreover, oblige potential consumers to take responsibility for their pain (whether through pill regimens or self-work) and do so by acknowledging that pain is both unnecessary and antithetical to human vitality. Placing the burden of responsibility on individual people in pain, rather than on public health, governments, and other social structures, has played a key role in shaping the opioid crisis into what it looks like today. For even today, as overdoses continue to rise, we have yet to confront the numerous social factors and social inequities that together constitute the broader context in which pain is experienced and lived.

DIY Pain Management

A deep dive into Amazon reveals a world of self-help products and literature that reinforce the obligation of individuals to take control of their pain—a world that is geared toward the practice of what I'll refer to as "do-it-yourself pain management." The do-it-yourself (DIY) approach to pain is promoted through various tools and techniques, all of which assert the need for pain sufferers to turn away from pharmaceutical treatments, surgeries, and expensive physical therapies toward products that enable them to create their own, individualized program. The options for what such a program might include abound: There are books and manuals brimming with at-home exercises, yoga techniques, and various forms of mindfulness training, along with all the necessary gear and equipment that such training might require. There are loads of pain-relieving vitamins, nutrition supplements, and diet plans. Outside of Amazon, the online world of DIY pain management is replete with YouTube videos, tutorials, and a host of websites and blogs that provide advice and reviews of the various programs, products, and physical/mental/spiritual exercises that are currently available on the DIY market. Many of these websites boast their own patented "methods" for achieving pain relief, each of which claims to be "backed by science" as an effective, drug-free approach to pain management. This chapter examines several of these texts, particularly those that were named Amazon bestsellers. These books remain enormously popular with their audiences, and each has earned both significant readership and far-reaching influence.

To take but one example, the website of "world leader in non-medical pain relief" Pete Egoscue is host to the owner's "Egoscue Method" along with links to his bestselling book titled *Pain Free: A Revolutionary Method for Stopping Chronic Pain*. Egoscue's "easy and gentle method," which reportedly "has a 94 percent success rate," is mostly made up of different kinds of calisthenics (or "motioncises") that "teach YOU how to regain control of your health without becoming dependent on another person or machine" (www.egoscue.com).

The Egoscue method highlights much of the rhetoric that is prevalent across the domain of self-help-oriented, DIY pain management. In this domain, pain management is driven by an ethos of individual autonomy: Consumers are hailed as being in control of their pain, as empowered, and as having a choice in whether they will continue to suffer. As a means of empowering people in pain, popular self-help titles encourage individuals to "conquer," "take charge," and otherwise discover, through a series of self-searching exercises and self-surveillance regimens, "the way out" of chronic pain.

In DIY pain management, to choose to remain a victim is anathema to recovery, while individual autonomy, in contrast, can be characterized in the mantra of "mind over matter." As Micki McGee has shown, "mind over matter" is a discourse that legitimates the promise of self-help, particularly as it relates to the body. The pull of self-help literature (whether diet and exercise manuals or methods for DIY pain management) lies in this vision of autonomy where "mind is master, body is slave . . . where the body is controlled by willpower or self-hypnosis" (McGee, 2005, p. 155).

The branding of pain relief in the self-help industry prizes individual autonomy and empowerment, values that can only be earned by training one's mind to become superior to one's body. This idea assumes that pain is as much "mental" or "emotional" as it is physical. As such, pain can be treated most effectively through various forms of self-work—in which the consumer invests in a product, book, or method to train their mind to acknowledge, confront, and eventually "let go" of pain. Take, for example, The Tapping Solution, one particularly popular DIY method for managing pain. The Tapping Solution is run by its CEO, Nick Ortner, who popularized the technique in a New York Times bestselling book, *The Tapping Solution: A Revolutionary System for Stress Free Living* (2014), and a documentary titled "The Tapping Solution" (2008), which provided the basis for the creation of five "Tapping World Summits" and various tapping-related charities (Ortner's Wikipedia page mentions ProjectLight: Rwanda, The Veterans Stress Project, and You Can Thrive).

As a therapeutic technique, the Tapping Solution is reportedly based on the theory of the "emotional freedom technique" (EFT), a counseling intervention that is derived from various theories in Eastern and alternative medicine, including that of Chinese acupuncture. EFT, like acupuncture, assumes that many disorders can be treated by redirecting and balancing the body's energy flows. The body's vital energy circulates through different channels, which in traditional Chinese medicine are sometimes referred to as meridians, which are connected to different bodily organs and functions. In acupuncture, needles are placed along specific meridians in order to correct an energy imbalance that is affecting the associated organs/functions. Similarly, EFT typically relies on placing pressure on meridian points. In a typical EFT session, a person will focus on the specific problem they are having, while tapping on various points on their body, often while repeating an affirmation or mantra.

YouTube contains a plethora of videos of different people demonstrating the "tapping" technique—some of which have amassed hundreds of thousands of views. These videos show that the Tapping Solution is, more or less, exactly what it sounds like: DIY pain management accomplished through the physical "tapping" of different body parts. If you are working with this technique, you may begin by tapping repetitively on your temples, then above your eyebrows, under your nose, and beneath your collarbone, all the while repeating various mantras. In one widely viewed video, Ortner himself demonstrates the technique using the following mantras: "Even though I have this pain in my body, I deeply and completely accept myself" and "I'm so frustrated by this pain, and I wonder if I can let it go." The assumption that underlies this method, and which is echoed across many others, is that by tuning in to one's emotional state, one can access the "truth" behind their pain.[1] As Ortner writes, "Your emotions are entry points into the core of whatever ails you. By burrowing deeper into how you feel, you're able to discover and unlock whatever it is that's holding you back and even hurting you" (2014). This paradigm reinforces an understanding, which is also part of the discourse in Purdue's "I Got My Life Back" videos, of pain as an obstacle to realizing and restoring one's authentic self. An individual who utilizes the Tapping Solution and finally makes the choice to "let go" of their pain achieves self-mastery over their body. Achieving this kind of mastery is, according to the paradigm, akin to freedom, and the individual who is free of their body is likewise empowered to access the "truth behind the pain."

The promise of wellness is not unique to DIY pain management but has been said to characterize the self-help industry as a whole. Colleen Derkatch has argued that the concept of wellness—an idea central to both Western pharmaceutically based medicine and the self-help industry—is perhaps a rather complicated (and somewhat insidious) one. It is, as Derkatch argues, predicated on "an entanglement of seemingly opposed logics of restoration and enhancement" (2018). That is to say, at the same time that people who use opioids seek to restore their bodies to a natural, normal, or free state

---

[1] See, for example, the DIY pain management method espoused in the book *The Mindbody Prescription: Healing the Body, Healing the Pain* by John E. Sarno, MD. Another popular pain management self-help text, the "Mindbody Prescription," takes a Freudian approach to pain relief, a kind of "talking cure" that understands pain as "rooted in repressed emotion." This method holds that recovery can only be achieved through one's "becoming conscious" of one's emotional state—which is articulated as the true source of pain.

("how I was before"), they end up doing so by constantly changing them—modifying them, enhancing them, optimizing them, essentially doing everything possible that excludes restoring them back to normal.

Similarly, Micki McGee (2005) analyzed the function of rhetoric of freedom within the self-help industry as one that is enrolled in the construction of "belabored selves." As McGee explains, the belabored self is a result of the industry's attempts to insulate its consumers from "affronts to [their bodies'] vulnerabilities through constant efforts at self-mastery and self-management" (2005, p. 174). These attempts, McGee suggests, are "futile" at best, for the belabored self can never be free. Freedom will remain out of reach because the "belabored self" is defined only in terms of its ability to labor—to be engaged in constant work, "not only at the office on one's professional tasks, but at home, on oneself" (p. 174). The DIY pain management industry, just as the pharmaceutical pain relief industry, is also involved in the making of belabored selves. For if pain is to remain at bay, one must constantly work to keep it that way. Just as in Purdue Pharma's branding of pain relief in its promotional videos, the branding of pain relief in DIY pain management belies the reality that freedom requires more work than the administration of a pill, or a program. What is actually required is an ongoing commitment to invest in that which enables a person to maintain control over their pain and to integrate an ongoing search for relief as a normal component of their daily life.

The point is, Colleen Derkatch says, quoting Carl Elliot, that what this cycle shows is that what we really want, and what we believe is possible, is that we can be not just well, but optimal. Better than normal. And what this means, then, is that wellness is not an end goal as it would seem but is an ongoing or, rather, never-ending practice of attempted self-regeneration. Perhaps unsurprisingly, Derkatch's argument in many ways echoes Joe Dumit's astute idea of chronic patienthood—that by taking pills to return to normal, we end up placing ourselves in an endless regimen of artificial enhancement.

But on another level, Dumit's and Derkatch's arguments resemble one another, first, because they show us that self-help—whether through pharmaceuticals or alternative routes—encourages us to become engaged within a system of ongoing self-government. Second, both arguments make explicit the astonishing irony of all this: that what the work of the rhetoric of returning to "normal" does is convince us to become voluntarily trapped in a never-ending process of self-modification, which is, as Derkatch points out, a "perpetually moving target" (2018). The discourse of chronic patienthood and the rhetoric of restoration/enhancement, as Dumit and Derkatch argue, form a kind of endless loop. That is, when argumentation from one of two logics seems exhausted (e.g., when chronic pain abates from an 8 to a 3), then the other kicks in because, well, one could always feel better. By cycling between the two logics (restoration and enhancement), the language of wellness circles back on itself.

Self-Help's Entrance into Pain Management

In 1998, two years following OxyContin's release onto the pharmaceutical market, a tremendous growth in self-help publishing and self-improvement culture took off. That year, self-help book sales were said to total some $581 million—an impressive feat considering that at the same time, publishing was being faced with otherwise declining sales and soaring return rates (McGee, 2005). The rise of self-help in American culture has been identified by some scholars as a phenomenon linked to broader cultural processes: the growing dominance of neoliberalism, an increasing emphasis on the self, and, relatedly, a myopic focus on the individual as the sole bearer of responsibility for achieving health, wealth, and happiness. These are themes that run through the chronic pain self-help techniques previously analyzed, but they also beg other questions, more closely related to this project: How, exactly, did self-help make its way into medicine, generally, and pain management, specifically? What kinds of medical theories linked to self-help logics drove this movement? And finally, what implications does the rise of self-help in pain management have for the politics of pain in the opioid epidemic?

The rise of the self-help industry for treating chronic pain is closely linked to the popularization of various theories and methods within the field of pain management. Since the 1980s, physicians in North America have been developing programs aimed at the self-management (rather than the medical management) of chronic pain and other chronic illnesses. One such program, developed by Kate Lorig of Stanford University, aims to bring self-management techniques to bear on chronic health problems such as arthritis, lung disease, heart disease, and stroke. Lorig's interventional program, which is referred to as "chronic disease self-management" (CDSMP), is based on the theory of self-efficacy, as espoused by psychologist Albert Bandura (1977, 1997). For Bandura, self-efficacy refers to the beliefs that individuals have in their own capacity to overcome problems and exert control over their own functioning as well as any external event that may impact their life. Put another way, self-efficacy is the confidence that one has in one's own ability to treat one's problems and obtain positive outcomes. In CDSMP, self-efficacy is key to the self-management of chronic conditions, since, theoretically, a belief in one's ability to implement behavioral changes should lead to more positive treatment outcomes—including better pain management and a diminished impact of chronic pain on the lives and day-to-day functioning of people who experience it (Lorig & Holman, 2003).

According to Lorig and Holman (2003), self-management is a necessary process in which those battling chronic medical conditions must participate since, "only the patient can be responsible for his or her day-to-day care over the length of the illness" (p. 1). To become a successful "self-manager," chronic pain patients—and patients with other kinds of long-term medical conditions—must engage in three essential "tasks," which include the proper medical management of their condition (taking medications, adhering to a special diet, etc.) as well as successful "role management" and "emotional management," which hinge on patients creating "meaningful behaviors or life roles" and "deal[ing] with the emotional sequeli of having a chronic condition." These tasks, moreover, occur in an outpatient setting, in patients' own homes.

Although Lorig's approach to self-management originated more than 40 years ago, the idea that pain patients are individually responsible for the day-to-day management of their condition remains popular today. In fact, the International Association for the Study of Pain (IASP) defines self-management as the "first rung of the ladder in pain care." Pain management in pain clinics, hospitals, or other inpatient settings, in contrast, takes a back seat. Rather than focusing on the social conditions of pain or on patients' need for social services to help them manage the day-to-day challenges they experience with chronic conditions, self-management aims to teach patients coping strategies that may enable them to "relearn how to live" with a chronic condition—of which they are the principal managers.

In the mid-1990s, Lorig's approach to the self-management of chronic conditions was taken up in Canada by Sandra Lefort, who reworked Lorig's model to function specifically for chronic pain patients. In Lefort's Chronic Pain Self-Management Program (CPSMP), participants with chronic pain conditions participate in a series of workshops facilitated by two leaders, one of whom must suffer from a long-term pain condition or be a close relative of someone who does. The program is highly structured, and leaders follow a manual to facilitate the workshops. Of the workshops' objectives, managing feelings of frustration, anger, and depression that often accompany chronic pain is central. Other themes include communicating with healthcare practitioners, managing medication consumption, and managing social isolation (Lefort et al., 1998). Workshop sessions typically involve lectures and exercises in light physical activity, relaxation, and communication, as well as the implementation of "action plans." In the CPSMP, "action plans" refer to a participant's decision to choose a goal and determine what kinds of activities need to be undertaken to achieve it. Plans are performed on a weekly basis, experiences with them are shared, and participants are then encouraged to help each other solve any problems that might be interfering in the development and realization of others' plans.

Lefort's CPSMP has been used for many years in Canada and was more recently taken up in studies on chronic pain management in Denmark. While researchers who have evaluated the program have indicated that it offers some new directions and possibilities for alternative ways of treating chronic pain (other than opioids), it remains unclear what kind of effect such interventions really have on chronic pain. In a randomized controlled trial conducted by Lefort and colleagues to test the efficacy of the CPSMP, short-term improvements were found among 102 participants in pain, dependency, vitality, aspects of role functioning, life satisfaction, and in self-efficacy. However, no significant improvement was found regarding the study's main outcome, pain-related disability (Lefort et al., 1998). Additionally, another pilot study with 45 adults found small effects on physical role function and pain intensity but found no effects in the main trial, which included 256 older patients with chronic pain who received nurse-led or psychologist-led CPSMP. Finally, a third and more recent study found that effects of the CPSMP on pain distress, disability, and pain beliefs in 141 older adults were somewhat maintained after one year. A Cochrane review of the effect of lay-led CPSMP saw modest, albeit short-term, improvements in participants' exercise habits, self-rated health, symptom management, and self-efficacy (Foster et al., 2007). But a randomized controlled trial of the Danish version found no significant improvements in pain disability and found only slight improvements in participants' emotional distress and illness worry (Mehlsen et al., 2017).

Although the results from these studies do seem to indicate potential for improving some of the sequelae that chronic pain patients experience, the effects of the CPSMP on pain itself are less clear. Even so, in 2015, the CPSMP workshop was reworked into a book written by Lefort, Lorig, and others. *Living a Healthy Life with Chronic Pain* channels aspects of the CPSMP into a text that reads much like the chronic pain self-help books previously discussed. In its introduction, *Living a Healthy Life with Chronic Pain* claims that it will teach the reader "to be a positive self-manager by being proactive about your pain" and asserts that its authors believe that "if you adopt this positive management style, you will live a healthier life."

The language of Lorig and Lefort's text reads like a business manual and follows a kind of business ethos: Throughout the book, patients in pain are referred to as "managers," and the work they must undertake to treat their pain outside of the doctor's office is cast in the language of doing "your job." In the beginning of the book, the authors suggest that, in fact, self-managing chronic pain is a "job" that is "much the same as any other manager" and includes "gather[ing] information and work[ing] with a consultant or team of consultants consisting of your physician and other health professionals." These professionals, as consultants, are responsible for "giv[ing] you their best advice," but, as the authors note, "it is up to you to follow through." "Following through" with a doctor's advice, moreover, means mobilizing common management strategies—tools that are central to any good business. These include "problem solving, making decisions, and taking action." And, if you perform well, "rewarding yourself."

As do other books in the chronic pain self-help genre, *Living a Healthy Life* channels practices gleaned from traditional Eastern medicine (relaxation, meditation, mindfulness, etc.) and combines them with techniques of self-surveillance that have become part and parcel of outpatient Western medicine (Armstrong, 1995). Patients in pain are encouraged to create activity and rest diaries, journal their pain symptoms, and otherwise keep track of the behaviors in which they are to be engaging as part of the CPSMP. Patients are thus incentivized to discipline themselves as part of their daily practice of chronic pain self-management. They are not only to engage in healthy activities, but also to maintain an ongoing record of their progress and to perform periodic evaluations of their diaries so that they may become more closely attuned to their symptoms, coping strategies, and adherence to their action plan.

One key difference that separates *Living a Healthy Life* from the kind of self-management applied in the CPSMP is the emphasis on pain's place within a community or larger social sphere. By and large, *Living a Healthy*

Life and other similar self-help guides on the topic of chronic pain do not attempt to situate pain patients within a community or support group. Instead, they are individually focused, offering solutions that each patient in pain should apply in their own life, and in accordance with their own individually tailored limits and goals. However, while such texts depart from the socially informed approach to pain that characterizes the CPSMP, they do take on the language that is typically found in community settings where pain is talked about and intervened upon. In particular, chronic pain self-help takes on the language of "recovery" typical of 12-step groups and embeds this language in treatment programs that, instead of relying on the support of a pain community, leave it up to the individual who experiences pain to engage in an ongoing practice of self-education, self-surveillance, self-treatment, and self-improvement. Illustrative of this language of recovery in chronic pain self-help are several popular texts, which boast titles such as *The Way Out* and *Pain Recovery: How to Find Balance and Reduce Suffering from Chronic Pain*. Other examples include *Recipe for Recovery: A Guide to the Twelve Steps of Chronic Pain Anonymous* and *Chronic Pain Rehabilitation: Pain Management That Helps You Get Back to the Life You Love*.

Texts such as these, as well as Lorig and Lefort's *Living a Healthy Life*, cast the management of chronic pain as a health problem that one must, if one is to achieve a good life, act individually to overcome. Patients, in this model, are encouraged to become experts on their illness, first through education and then through various forms of surveillance/tracking. One example of this can be found in the self-help guide *The Way Out* (2021), which advocates for patients engaging in what the author refers to as "somatic tracking." Somatic tracking, according to this text, is a mindfulness technique akin to medication, through which a pain patient learns to identify, explore, and accept their pain sensations with "curiosity and levity" rather than with "judgement and intensity." In the process, somatic tracking is intended to "change your brain's relationships with pain" (p. 116). The implication, here, is that pain is rooted in the individual's brain, and that by engaging in disciplinary practices that reroute mental processes, an individual may successfully manage chronic pain without the use of opioids or other kinds of pain medications. While this idea has some merit in that it offers patients an alternative avenue for exploring relief, it also places pain patients in a potentially uncomfortable situation, one where they have become responsible not only for the treatment of their pain, but also for their treatment *outcomes*. That is, patients whose pain does not improve following self-management interventions are not seen in terms of the broader social and economic conditions that may be impacting their lives and treatment. Instead, they are seen as ineffective self-managers.

While individually focused approaches to pain such as those found in self-help programs can be seen as offering new opportunities to pain patients by empowering them to become more educated and therefore more adept to manage their conditions, they also risk missing out on pain's bigger picture: that is, such an approach to chronic pain ignores the systemic problems that condition the prevalence of chronic pain in North America—for example, the working conditions of the working poor, the systemic lack of access to healthcare in working-class and poor regions across the country and throughout the life span, and the widespread social isolation that often contributes to chronic pain symptoms and renders existing treatments ineffective (Craig et al., 2020). Additionally, the self-management approach to chronic pain, when bound up with individual patients and their responsibility to improve or "recover," reifies self-surveillance as a necessary component of a pain-free life, a practice that runs counter to the empowerment narrative that dominates chronic pain self-help. As Isabelle Baszanger (1998) has written in *Inventing Pain Medicine*, the approach to chronic pain that such self-help models take is one that attempts to discipline and correct not pain itself but, rather, the poorly adapted behaviors that lead one to experience unlivable chronic pain. Yet, even in this more holistic behavioral model, the chronic pain patient is not necessarily any more empowered than the patient who takes opioids to treat their chronic illness.

The Pain Patient-Expert

Self-help's ascendance in medicine, especially for chronic conditions like pain, is paralleled by what Joy Fuqua (2012) has referred to as the rise of the consumer-patient—and what I refer to as the rise of the *pain patient-expert*. The patient as a consumer is a concept that is linked to several trends taking place in medicine throughout the 1970s and 1980s—including the deinstitutionalization of medicine, the growth of the medical marketplace, the development of direct-to-consumer drug advertising, and others—which paralleled the reconceptualization of the patient's role in healthcare. As Fuqua writes, prior to the 1980s, patients were often conceived as passive receptacles of treatment, and as individuals who held little power when it came to making diagnoses and developing treatment plans. Yet with the entrance of market-based language into healthcare, patienthood was rewritten, and patients became envisioned not as passive recipients of care but as "active, information-seeking" consumer-patients (Fuqua, 2012). Such a reconceptualization, as Fuqua notes, suggests an evolution in patienthood and the development of a new kind of patient-centered freedom that rejects older, paternalistic models of care. And certainly, to some extent, the patient-consumer is free—free to choose their treatments. However, it is worth noting that new freedoms, in this case, also carry with them new responsibilities. For if patients have become the principal agents of medical care, who are free to select and consume medical treatments, then they are also responsible for the effects of those treatments—for their success or, alternatively, their failure.

The rise of the patient-consumer is closely linked to the rise of the patient-expert in pain management, with one key difference: While patient-consumers are granted the ability to choose their treatments, patient-experts are further endowed with the ability (and responsibility) not only to choose but also to *perform* these treatments, which may or may not lead to positive outcomes. One example of the responsibilization of people in pain can be found in the book *Pain Recovery: How to Find Balance and Reduce Suffering from Chronic Pain*, which presents an "opioid-free approach" designed for chronic pain patients who have or currently take opioid pain medication. In this book, like many others about chronic pain, the patient is encouraged to take steps to regain control (which is coded in *Pain Recovery* as "balance") of their bodies, their minds, and, as a result, their pain. *Pain Recovery* proposes to work as a guide for patients to do this by offering chapter sections that elaborate on the importance of "self-talk," "evaluating your pain," and "maintaining physical balance." While *Pain Recovery* does, to some extent, recognize that pain is multifaceted, it still insists on providing recommendations that locate chronic pain "recovery" within individuals and their willingness to become active agents who are responsible for restoring the quality of their lives. Like the other self-help texts discussed in these pages, *Pain Recovery*'s self-help discourse positions the pain patient as a new kind of pain expert, who, from experiencing pain, is best suited to manage it. To that end, self-help texts propose mindfulness and individual action (such as self-education about pain processes) as individual solutions to the social problem of chronic pain.

Like Fuqua's patient-consumer, the pain patient-expert connotes a specific kind of process, not one in which passive patients evolved into active experts but, rather, one in which patienthood and expertise have become combined and conflated. What conflation means here is that the pain patient-expert's newfound "freedoms" are not entirely without their limits. On the contrary, the patient-expert is subject to many limits, which restrict unbridled medical freedom. First are the limits that apply to pain patient-experts who cannot actively choose their own healthcare provider, whether for financial, geographical, or other reasons, or who cannot afford to pay for health insurance that promises access to all available treatments. Second, in a larger sense, the patient-expert is a subject of capitalism and the medical marketplace. As such, they are only free to choose among a limited number of medical products, which are dictated by the ebbs and flows of the medical market. Thus, like the patient-consumer, the patient-expert's newfound freedom is severely limited by market logics.

The rise of the pain patient-expert can therefore be seen as a reflection not of expansions of freedom in medicine but, rather, of increasing limits on that freedom. The balance of power among patients and providers is revealed to be unequal, since ultimately, it is the provider, insurance company, and medical marketplace that decide who gets what treatment, what it will cost, and for how long it will be available. Even so, the onus of responsibility for positive outcomes remains with the patient.

The focus of pain self-management on the individual person and their behavior is significant, especially in a context where opioids are—for many people in pain—difficult to obtain. What this means is that pain undertreatment has become a constant problem for pain management and, more importantly, for people who experience chronic pain. Since the start of the opioid epidemic, prescribers and health institutions have been developing and releasing guidelines that attempt to limit the prescription of opioid painkillers. These new limitations on prescribing augment the difficulty for individuals seeking relief from chronic pain, who may also be struggling with the financial costs of opioids (especially if they do not have insurance) or with the problems of navigating a regulatory landscape in which these drugs are tightly controlled. This problem space is where pain self-management and non-drug self-help have stepped in, since they offer additional market-based solutions for people suffering from chronic pain.

The Biopolitics of Chronic Pain Self-Help

The reliance on market-based solutions for the problem of pain has led to the construction of people in pain as market-based subjects, whether they are situated within or outside of the pharmaceutical industry. Through chronic pain self-management, the person in pain is envisioned—and encouraged—to be, in Foucault's terms, an "entrepreneur of himself" (Foucault, 2008). This is the biopolitics of chronic pain self-help, the political process through which the bodies of individuals who experience chronic pain are controlled and made useful (Fernandes et al., 2016). Moreover, the mobilization of the biopolitics of chronic pain takes place primarily through an emphasis on individual freedom, choice, empowerment, and agency in one's own recovery. Yet this approach to pain is fundamentally incompatible with the reality that pain is—and must be seen as—a social phenomenon. The experience of pain and the possibilities that exist for its treatment are different for members of different social groups. Well-resourced groups are, in general, better able to manage their pain than are members of disadvantaged groups. This is not because of any inherent difference in individual ability, but because of deep-seated inequalities in healthcare access and treatment options, which have been shown by several recent studies (e.g., Craig et al., 2020; Wallace et al., 2021; Ziadni et al., 2020) to be severely limited for historically and persistently excluded groups. For example, one recent study in the United States showed that participants who experienced more social injustice also had much higher pain intensity than white participants who did not (Trost et al., 2019). Thus, managing the pain of these individuals has proved to be a much more complicated and difficult task than it has been for individuals who do not experience social injustices and who are, more often than not, white and well resourced.

What this and other studies go on to show is that pain is deeply entangled with and shaped by social locations and social identities. It also shows that social context is essential for both understanding and treating chronic pain. Yet, this context is often missing from both pharmaceutical and nonpharmaceutical approaches to managing it. In chronic pain self-management discourse, what we see is not an attempt to disentangle the complex experience of chronic pain from its social roots but, rather, a myopic focus on individual wellness as another technological fix for chronic pain.

But why? It's important to note that the promotion of technological fixes—whether they are pharmaceutical or not—in self-help discourse is strategic.

By focusing on individuals and the problems that they are now responsible for solving, self-management discourse avoids rethinking the organization of social institutions and their priorities, a step that would be as problematic for a profit-hungry pain industry as it would be beneficial for people in pain. As a result, what we have seen in the opioid epidemic, and what we are likely to continue seeing in the future, is the ongoing construction of "empowered," responsible, entrepreneurial subjects in pain—whose suffering is addressed not by socially informed solutions, but by entrapment within a never-ending cycle of consumption in the name of self-restoration and self-enhancement.

So, what's so different about the logics and rhetoric of these pharmaceutical and nonpharmaceutical solutions to pain? The answer is: very little, if anything. We still see at play the promise of freedom from pain, the guarantee of self-restoration, and the harsher and more complicated realities of chronic patienthood. The primary vehicle that drives the advancement of this rhetoric, these logics, these seductive dreams of absolute freedom from pain, self-restoration, optimization, and the realization of one's American dream is branding. The branding of pain relief is both an optimal medium for rhetorical construction and a formidable process of emotional production and meaning making. It is not all powerful, and it is not the only driving force behind this epidemic, but it is one that we cannot ignore. For it is this mechanism that has been pushing this particular hamster wheel, which, as it furiously rotates and cycles round in its closed system, tries and succeeds to incorporate more and more of us every day.

* * *

In 2014, the Pew Research center released a survey report titled "America's New Drug Policy Landscape." What the survey showed was that the American public is, by and large, ready for a dramatic shift in drug policy and the methods that have typically been used in America's waging of its three-decade long "war on drugs." Of the respondents, 67% said the government should focus less on the typical criminal justice approach to drug use and the prosecution of drug offenses and more toward the treatment and rehabilitation of PWUD. Only 26% thought the government should focus more on its prosecution of PWUD. Local law enforcement agencies around the country are also attempting to reorient their approach to drug criminalization—namely, by distinguishing those who *use* drugs from those who *deal* drugs, and by prosecuting only the latter group.

Yet, the gray market for opioids presents an enormous challenge to regulators and lawmakers attempting to flesh out these distinctions, as it is becoming increasingly difficult to identify who exactly is a "user" and who is a "dealer." If my grandmother gives my father, her son, a half-used bottle of OxyContin for his lower back pain, is she now considered a drug dealer? Depending on the quantity of pills she handed over, the answer could actually be yes. By some other standard (perhaps a legal standard of "intent"), the answer would be no. And what about physicians? Who is to say whether prescribing a month's supply of an addictive opioid is "reasonable" within the standards of medical practice? It could just as easily, depending on what that prescription amounted to (whether or not the patient became habituated to it, overdosed, etc.), be interpreted as malpractice, manslaughter, or even murder (as is now sometimes the case) (Cohen, 2016; Sullivan, 2017; Wootson Jr., 2017).

These are some of the questions that concern us in the next chapter, which focuses on the ways in which the opioid epidemic is fundamentally a gray matter—defined by uncertainty, ignorance, strategies, and powerful rhetorical and cultural devices that have shaped its trajectory. In particular, this chapter contends with the rhetorical strategies and concrete practices that have evolved in the face of the grayness of pain—how, while many old rhetorical boundaries have been dismantling, new ones are being erected, with profound effects on the subjectivity of those whom they attempt to define. The rhetorical boundaries that are erected in the face of uncertainties and fears surrounding painkiller use are mobilized through discourses of victimhood, class, and race—all of which are key elements that have given shape to the opioid epidemic and address the question of whose pain matters and why.

5
Pain's New Faces

As we have seen, in medical, regulatory, and cultural environments alike, certain boundaries and "truths" about opioid use and people in pain have begun to erode. Narratives of the opioid epidemic are brimming with evidence of categorical breakdowns, in particular where patients become easily framed as addicts, where doctors become prosecuted and sentenced as dealers, and where cures become seen as poisons. All around us, it seems like solutions, U.S. Food and Drug Administration (FDA) drug research efforts, drug companies' "education" campaigns, the pain self-help industry, and so on—try as they may to provide solutions to the crisis—continue giving rise to new and greater problems. Specifically, each of the semantic shifts listed above reveals another systemic issue—a crack in the larger classificatory system of public health. As Foucault and Burke (among other philosophers and rhetoricians) have discussed, when this kind of threat rears its head to long-held definitional systems, the urgent need to reclassify and re-establish meaningful distinctions becomes even more acute. That being the case, a study of the opioid epidemic would be incomplete without a focus on the ways in which these breakdowns are affecting and being affected by various discursive attempts to restore order to what now appears to be a rhetorical and institutional system in crisis. These are primarily carried out by the discursive construction of people who use opioids as different kinds of persons—persons who I define in a typology that consists of "patients-at-risk," "pathological consumers," and "consumed addict-victims."

The rhetorical-cultural analysis of pain that follows entails looking closely at the ways in which subjects of pain are framed—particularly when it comes to opioid use. But this process of subjectification stretches beyond the individual person in pain themselves; it also feeds into much longer historical and cultural tendencies—which we will review, and which are reflected in politics, pop culture, etc.—that say something about the relationship between opioid use, people who use opioids, and the American dream. The construction of people who use opioids endows value upon certain kinds of pain-inflicted subjects, as well as defining who they are, what we should expect them to look or behave like, and what their relation is and should be to pain and addiction and, by extension, to the crisis as a whole.

Ever since the first reports of prescription opioid-related overdoses began surfacing in local newspapers in 2000, the mainstream media have been publishing stories trying to understand what the "opioid epidemic" really is—what are the drugs, who are the perpetrators, who are the victims, and so on. One of the most common queries, which lies at the heart of the epidemic's construction in popular culture, attempts to describe what are often characterized as "the new faces" of the opioid crisis. Consider, for example, a piece the *New York Times* ran in the summer of 2001, just after the first local reports of OxyContin overdoses: The piece, which is dramatically titled "The Alchemy of OxyContin," is narrated by a reporter who takes a midnight drive with Paula, his informant and a self-identified OxyContin addict living in a small town in West Virginia (Tough, 2001). As they navigate the city, Paula casually points to the comfortable-looking homes and neatly parked cars belonging to the local drug dealers: nurses, soccer moms, and the high school prom king, none of whom seemed to resemble the "kind of person" one would expect to find dealing narcotics on the city's darkened street corners. And yet, as the report suggests, this is the alchemy of OxyContin—a drug with the ability to fashion *new* kinds of persons, transforming them seemingly overnight: Suddenly, patients transmogrify into addicts, soccer moms into felons, prom kings into corpses.

Or consider another story, this one an ABC News segment, which aired in October 2010: "The New Face of Heroin Addiction" followed the stories of several opioid-addicted individuals living in suburban areas of New England (Whitaker, 2015). What was "new" about the faces of the epidemic, ABC host David Muir shared with viewers, is that they belonged to "kids from families you never thought could fall victim to drugs ... families who did everything right and still their kids are hooked." A 2015 CBS segment titled "Heroin in the Heartland" was similarly focused on the surprising "new look" of the epidemic—"the faces of heroin," which are "young, middle-to-upper class and suburban." What opioid addiction's "new look" tells us, and what each of these news segments implies, is that opioid use is no longer the province of "criminals." Nor is it any longer an inner-city problem. On the contrary, it seems to suggest that opioid addiction has become something that can just happen—and to anyone. As one of CBS's informants exclaimed, "Even Miss America could be a junkie!" By belonging to anyone, this "new face of addiction" helps make the opioid crisis intelligible as a nationwide epidemic, one that is all of our problem and that we are all called upon to care about and attempt to resolve.

In more analytical terms, what these narratives illustrate is the development of a process that is centrally involved in the rhetorical construction of the U.S. opioid crisis—especially as it concerns the categorization of its subjects. As I see it, the work these representations ultimately accomplish is to divide people who use opioids into (at least) three categories of persons: First is the patient-at-risk. The risk, which is always present in discourse about opioids, refers to addiction and the ever-present danger of an innocent patient being transformed into an opioid addict. But opioid "addicts," too, are not all one and the same. On the contrary, opioid-addicted subjects are constantly being produced on either side of a division: On one side stands a familiar subject, the addict of history, a dangerous individual whose addiction is understood in terms of their status as a pathological consumer. The addict as pathological consumer is, in the opioid epidemic, supplemented by the construction of third kind of addict who stands on the opposite side of the divide: Unlike the pathological consumer, this other addict is not a danger but is instead endangered. They are not consumers but are instead positioned as victims of malevolent forces outside themselves, forces that have consumed them.

The production of these three categories of persons—of patients-at-risk, pathological consumers, and consumed addict-victims—constitutes one of the main thrusts of classification in the opioid epidemic. And, as the text that follows argues, subject production is critical not least because it serves to reinforce ideas about whose pain really matters, and whose does not. That is, classification actively shapes the meanings associated with human pain. And in doing so, these definitions provide a strong rhetorical basis upon which to argue on behalf of certain kinds of people who use opioids, to advocate for those who are worthy of saving, while proposing alternative strategies or justifying inaction on behalf of others who are seen as beyond rescue, or less worthy of our help.

* * *

In a critical discourse analysis of the representations of opioid use in popular culture, medicine, and policy, this chapter focuses on the ways that people who experience pain, especially when they use opioids, have been selectively constructed as different kinds of persons. These distinctions, moreover, have emerged in the wake of historical representations of some people who use opioids as patients or victims and others as addicts or criminals. These representations span back many decades but can also be shown to relate to the recent push in politics and public discourse toward a compassionate approach to drug enforcement for some patients, but not for others. For those whose opioid use warrants compassion, addiction is repositioned away from the domain of the criminal into the calmer, cleaner realm of public health. And it is within this public health framework that some people in pain are made legible. While not entirely new, the move to redefine opioid use as a public health issue rather than a criminal justice issue goes against longstanding trends in the history of the United States' "war on drugs" and the ways in which it has systematically policed, criminalized, and incarcerated people who use drugs—namely, poor people of color (e.g., Alexander, 2010/2012; Marez, 2004).

Thus, opioid use today is not only characterized by discourses of compassion and public health but also by a key shift—one in which people who use opioids who were formerly defined in terms of their criminality are supplemented (and in some cases, replaced) by the figure of a person in pain who is defined in terms of their patienthood. This process is likewise bolstered by the outpouring of stories over the past few years about people who use opioids, whose addiction trajectory began not on the streets with a needle in hand but with a trip to the pharmacy and the innocent purchase of an opioid medication.

One of the earlier and most infamous examples of this narrative can be found in the waves of media attention that followed the sudden death of acclaimed actor Phillip Seymour Hoffman in 2014 and the widespread fears and concerns they raised about the growing phenomenon of the "white prescription opioid user-turned heroin addict" (e.g., Bernstein, 2015; Calabressi, 2015; Gusovsky, 2015; Lavitt, 2015). Seymour's death formed part of what would become a nationwide moral panic surrounding the unexpected risks—particularly for suburban and rural teenagers and young adults who had been prescribed narcotics for sports injuries—of becoming one of the "new faces" of American drug addiction. Media coverage of this "new epidemic" is illustrated in across headlining articles and press coverage (e.g., Schwartz, 2012; Carroll, 2014). These articles, in highlighting the tendency for "normal," "upstanding" kids, teachers, and other professionals to become "hooked on drugs," draw an implicit contrast to a population of people who use opioids—a population that has been, up until now, the focus of most drug epidemics. That is to say that the "new face of drug use," which lacks the marks of race and ethnicity typically associated with urban people who use drugs (PWUD), lends a new sense of urgency to this problem—one that also undergirds the decision made by then-President Trump to declare the escalating rates of opioid abuse a national public health emergency.

Today, the person who uses opioids—who is positioned as both white and as a subject within the realm of public health—is not as problematic a figure as the "addicts" of history. That is, today's opioid consumer, particularly if they are white, is not positioned as a criminal, or a deviant, or a threat to the social system. On the contrary, they are exemplary of the subjects in pain who are the new faces of the epidemic: These subjects are *pain patients-at-risk*, citizens who, by some error in the course of their participation within the medical system, have been transformed into its victims. The pain patient-at-risk is at the center of what appears to be a new kind of struggle against drug use, a "white war on drugs," in which blame is repositioned away from certain people who use drugs and attributed to other forces external to them.[1]

The discursive construction of the pain patient-at-risk is likewise bound up with the question of how to distinguish legitimate pain patients from illegitimate opioid addicts. Until now, the answer to this question has been seemingly straightforward: The distinction has been based, in large part, on the specific methods that people in pain employ in their consumption of opioids. That is to say, the consumption practices of people who use opioids are deemed valid so long as they appear to align with the specific medical, legal, social, and ethical values and norms that govern the medical standard of taking drugs "as prescribed." Those who consume opioids as prescribed have long escaped the judgment and scrutiny of those who chose instead to take the other path—the path of "abuse." These people are not seen as patients but as addicts, with their consumption practices being defined as such on the basis that they appear to deviate from this standard.

The rule of "as prescribed," as has been discussed, is also tied to a legal logic, which has protected medical practitioners, regulators, and pharmaceutical companies from prosecution in cases where an individual has been harmed or died after taking a medication. After all, neither a doctor nor an FDA representative nor a drug manufacturer can be held liable for a consumer's own

---

[1] It is worth noting that the compassionate approach to opioid use and people who use opioids is not all encompassing. There are exceptions to this approach, which can be seen in other tendencies within the larger cultural and political landscape that surrounds opioid use. One of these tendencies has to do with the widespread reluctance to loosen regulations and prescribing guidelines on Narcan and other life-saving medications.

willful misconduct. If that consumer did not take their drug as prescribed, it is they—and they alone—who must bear the consequences. Yet, in the past few decades, as more and more individuals have suffered and died from taking their pain medications, this logic has been called into question. The oft repeated story of the opioid patient-cum-addict, of an individual who has become addicted, overdosed, or perhaps died, after taking their opioids exactly as they were told to do, exposes a crack in the logic of the golden rule of "as prescribed." The possibility that even the most obedient patient may become, unwittingly, an opioid addict poses a serious threat to the rule's legitimacy. And in the world of prescription opioid use, it has already broken down. In its place has emerged the possibility for seeing all opioid patients as potential addicts, or pain patients-at-risk.

The pain patient-at-risk is important not only because they are a salient character that has contributed to the intelligibility of the epidemic, but because they occupy a key position in many of the legal arguments that are currently being waged against opioid manufacturers and distributors. At the time of this writing, upward of 20 different states and thousands of cities and counties have filed lawsuits suing opioid manufacturers, including Janssen Pharmaceuticals, Purdue Pharma (the maker of OxyContin), and Endo Pharmaceuticals, the manufacturer of the drug Opana ER, which was recently pulled from the market. Lawsuits against these various corporate actors have been piling up over the past decade, including that which led to several historic, multimillion, or, in the case of Purdue Pharma, billion-dollar settlements and the latter's desperate attempt to file for bankruptcy. But in addition to these cases, some of which are now settled and some of which are still in progress, investigations have been launched by upward of 40 different state attorneys general, all of whom are attempting to gather evidence to pursue future litigation. Centrally involved in the attorneys general cases against the pharmaceutical companies is the collapsing distinction between opioid patients and opioid addicts. That is to say, the legal war they are waging is waged on the basis that people who use opioids do not overdose or become addicted as a direct result of misusing opioid products but that "they do so when used as directed, too" (Deprez & Barrett, 2017). The year 2003 saw a cluster of lawsuits taken up against one manufacturer in particular—who has by now become to readers a rather familiar character: Purdue Pharma was accused in these lawsuits, namely, by the families and friends of overdose victims, of knowingly pushing drugs on their loved ones, each of whom eventually became addicted to them "despite obtaining them legally and taking them as directed."[2]

While pharmaceutical companies have certainly found themselves facing an enormous legal problem, one founded on the production of the pain patient-at-risk, this subject also poses a threat for other actors. Increasingly, the accused parties are also physicians who are the defendants (some already imprisoned) as a result of several ongoing government lawsuits. Regardless of whether it is fair to lay the blame at their feet, it is certainly the case that doctors represent one of the groups that have a stake in maintaining the distinction between patients and addicts as it exists in the rule of "as prescribed." For it is now more often the physicians rather than the pain patients who are held responsible when an individual dies from taking their medicine. Other actors with stakes in maintaining the distinction between "as prescribed/not as prescribed" and patients/addicts include regulators, insurance companies, and chronic pain patients: Regulatory agencies (as well as pharmaceutical companies), who were previously shielded by the assumption that addiction would not befall opioid consumers unless they took their medications in a way other than as directed, must now spend additional time, effort, and money on risk management programs. So too, chronic pain patients and advocates have also had a stake in maintaining the "as prescribed/not as prescribed" distinction. For it is this distinction that underlies many of the arguments currently being waged by pain advocate associations against the government's implementation of opioid prescription limits. These changes, by making more difficult individuals' access to the medications they need to live normal lives, have thousands of pain patients rising up in anger and fear of a population of reckless consumers whose behavior threatens their future access to medications (e.g., Frostenson, 2017; Kertesz & Gordon, 2017; Kertesz & Satel, 2017; Keilman, 2017). Thus, the golden rule of "as prescribed" no longer works as it is supposed to do—to distinguish patients from addicts as different kinds of legal subjects. Even more, its failure undercuts the assumption that there is an inherent difference between them. Despite this (or perhaps because of it) the pain patient-at-risk is constructed in ways to remain distinguishable from the figure of the opioid addict.

---

[2] Whether as a direct or indirect result of these charges, Purdue Pharma recently announced that it will stop promoting its opioid products to physicians (Poston, 2018)—a rather uncommon course of action that indicates a drastic shift in the world of drug use, where blame is repositioned from PWUD to external actors and institutions operating within the medical–industrial apparatus.

Classifying People Who Use Opioids: Consumers and the Consumed

In the opioid epidemic, the figure of the person in pain who uses opioids is further split into two categories: the person who uses opioids as a pathological consumer, on the one hand, and the person who uses opioids as a consumed victim, on the other. But before examining these categories in more detail, it's necessary to step back to discuss the concept of opioid use and its historical evolution. For opioid use is a term with a complicated historical trajectory. Tracing the development of the concept of opioid use—and therefore of the person in pain who uses opioids—is crucial if we are to understand how these two subject categories could have come into being in the first place.

In the opioid epidemic, the discursive construction of opioid use and of people who use opioids is likewise concerned with behaviors that are seen as being reckless and destructive. Yet the source of the person who uses opioids' destruction is not seen as being the same for all people who use opioids. On the contrary, as this section details, classification in the opioid crisis divides people who use opioids into two different kinds of persons, one for whom destruction is squarely located within that individual and their behavior and another for whom destruction comes from forces acting outside themselves, which they cannot always hope to control. Thus, the world of opioid use divides its subjects into the following categories: pathological consumers (not so different from the "dope fiends" or the "junkies" of old) on the one hand, and those who are consumed addict-victims, on the other.

And yet, discourses about opioid use and addiction that divide people who use opioids into those who are to blame and those who are blameless are not themselves anything new. On the contrary, as others have discussed (e.g., Alexander, 2010/2012; Conrad & Schneider, 1992; Lassiter, 2015; Bourgois & Schonberg, 2009; Jacques & Wright, 2015), the construction of opioid use has long divided "addicts" into binary categories of good and bad—categories that have often mapped directly onto racial divides. The system of racial relations within which PWUD have been segregated and classified as persons who are either good (the white patient) or bad (the Black addict), and which map onto the treatment of opioid use as a public health crisis or a criminal justice issue, is somewhat complicated by the emergence of the pain patient-at-risk and with the dissolution of the rule of "as prescribed." Both processes have been crucial for distinguishing—legally, institutionally, and culturally—the different kinds of opioid consumption that belong to pain patients and people who use opioids. Yet, in the face of such ambiguity, the discursive construction of people who use opioids as being either consumers or consumed should also be understood as a strategy—one in which various actors have participated as an attempt to restore order and grant legitimacy to the previous system of classification. But what is ultimately key to take away from this typology—especially from the two following categories—is the key cultural and political work they do: For what they do is, on the one hand, justify the use of a "compassion" based discourse and treatment regimen among some people who use opioids, those who are seen as victims of a larger public health crisis. On the other hand, those who are not seen as consumed victims but, rather, as pathological consumers, are not so easily given the benefit of the doubt of compassion.

The person who uses opioids who is seen as a pathological consumer is defined on the basis of their individual behaviors and social role—both of which are seen as deviant, morally dubious, and as manifestations of an underlying pathology or character deficit. And opioid addiction, for the pathological consumer, is understood in terms of that person. That is, it is understood in terms of the personality traits, family history, and other clinically articulated "predispositions" that make them more susceptible to deviant behaviors and anticipate their tendency toward future substance abuse (Carr, 2013). Their behaviors in these domains are furthermore understood as forms of willful misconduct (Valverde, 1998). They are intentional actions, ones that position the person in pain not only as a pathological opioid consumer but also as a "hustler"—an active participant who knowingly exacerbates (and to some extent, thrives on) their transgressions. As a hustler, the pathological opioid consumer's condition is also associated with a range of other behaviors, which may include stealing, pimping, whoring, pushing, and so on (Bourgois & Schonberg, 2009; Knight, 2015). Thus, the pathological consumer is, in many ways, the addicted subject of history, one that closely resembles the character of the heroin junkie.

To provide a visual example, Figure 5.1 contains a photograph that illustrates many of the characteristics of the subject of the pathological opioid consumer—as well as the relationship they have to their drug of choice. This image, which was posted on the website of White Sands Treatment, a rehabilitation center in Florida, depicts a young, white, female opioid user, who is centered in the frame against a black background. She

Figure 5.1. Stock photo depicting an opioid "addict" used on the White Sands Treatment website.

Note: Stock photo—purchased standard license agreement (reproduction up to 500,000 copies). Accessed 10/20/22. Reproduced stock photograph found on the following website: http://www.whitesandstreatment.com

stares directly into the camera. In her left hand is a needle, full of brownish fluid, which she grips between her middle and ring fingers. She is showing off, performing her addiction with a maniacal grin and coy squinting eyes as she proudly displays her paraphernalia to whoever might be watching her. Her hair is messy and her eyes slightly out of focus, signaling to the viewer that she just had or is about to get her fix and that she is in no way ashamed of it. This person, this pathological consumer, does not give any hint that she intends to recover. In fact, she seems to suggest the opposite. She is many things but treatable or rehabilitatable she is not. After all, the photo implies, she has agency, a choice of whether to engage in problematic behavior—and she has already made it.

On its website, White Sands Treatment also depicts another kind of opioid user, who appears to be the same girl but whose photo is sharply juxtaposed to the one above. Figure 5.2 provides a visual example of opioid use as it manifests in a person who is not an active, willfully pathological consumer but is instead depicted as passive and remorseful—an addict-victim who

Figure 5.2. Stock photo depicting an opioid "addict" used on the White Sands Treatment website.

Note: Stock photo—purchased standard license agreement (reproduction up to 500,000 copies). Accessed 10/20/22. Reproduced stock photograph found on the following website: http://www.whitesandstreatment.com

does not want to consume the drugs but has already been consumed by them. She is, as her offset gaze would indicate, under the influence. But this is not in the sense that she is high but rather in the sense that she is brainwashed, a prisoner bound to forces that (although they may be small or even unseen) hold a power that is much larger than that which she holds herself. Unlike the larger-than-life pathological consumer, the person in pain who is consumed is even smaller and less powerful than the tiny pill she grips between her fingers. She is not a performer but looks shy and frightened, like someone who would rather be anywhere else in the world than where she is right now. Her face is tear-stained and sad. Her eyes gaze sideways, cautiously considering the orange capsule in front of her. But this pill does not make her happy. In fact, she seems to resent it. It's the pill's fault, she seems to say. If she could, she would throw it away. But she cannot. Her feelings toward it are complex: She wants it, but she doesn't; she loves it, but she hates it; she is scared of it, but she still can't take her eyes off it. Even as she grips it, she is even more so gripped by it—not a consumer but a victim who has already been consumed.

The discursive construction of people who use opioids—these "new faces" of people in pain—repositions the moral space of opioid use and addiction away from individuals toward the institutional arrangements that produce and consume them. Seen from the vantage of the consumed addict-victim, the story of opioid use "is not a story about dark alleys and drug dealers" but, rather, the story of something that "starts in doctors' offices" and "with everyday people seeking relief from pain and suffering" (Calabressi, 2015, p. 28). This repositioning of opioid use from the street into these new spaces also coincides with a repositioning of blame, which shifts from the individual themselves to other forces external to them. In this sense, the concept of opioid use as it relates to the consumed subject represents a departure from the other ways in which it has historically been understood. For the opioid addict-victim, opioid use and addiction are not the disorders that underlie pathological consumption but are instead a process of becoming consumed. What people who use opioids are consumed by, moreover, is not limited to a drug but also includes all the forces that constitute the medical–industrial complex—big pharma, physicians, insurance companies, and the regulatory apparatus are all seen as having been culpable parties in the process by which so many Americans have become habituated to opioids.

Defining People in Pain

While the categories of people in pain who use opioids as either consumers or consumed mark a crucial characteristic of the opioid epidemic, it is not the case that this kind of meaning making is just now taking place. On the contrary, there exists a long and complicated tradition of dividing people who use opioids and PWUD into different categories, which lends historical footing to the typology of people in pain as patients versus addicts. Distinguishing pain patients from opioid addicts and, in doing so, deciding whose pain has been and is being made to matter could not be more central to the story of the opioid crisis.

Throughout the 19th century and first two decades of the 20th century, American attitudes toward opioid use—and toward opioid addiction—were marked by ambivalence. In the 19th century, especially, opioid use was not considered a major social problem, namely, because the opioid drug morphine was widely available and used as a medicine for a variety of symptoms; as such, it was typically associated with its medical uses and was largely seen as something favorable, rather than a potentially addictive and dangerous narcotic (Conrad & Schneider, 1980). Correspondingly, individuals who were habitual consumers of morphine were not as readily stigmatized as addicts and tended to arouse sympathy rather than contempt.

While it was already known that addiction to opioids was possible (and that it already existed among people who used them), the condition itself nonetheless did not spur such moral outrage as it eventually came to, in large part because it was seen as a kind of sickness. This sickness, "morphism," understood as a disease not unlike any other, was suffered by ill people who were pitied as victims of their condition (Courtwright, 2001). The victim status of people who use opioids during this period was also associated with the race and class of morphine consumers—the majority of whom were white, middle-class, and middle-aged individuals living outside of the city centers (Keier, 1988). A large percentage of morphine-addicted individuals, moreover, were women, many of whom received the drug from their physicians as a means of alleviating pain associated with menstruation and other similarly "female" problems. For them and others, morphine not only was seen to work to taper physical pain but also was prescribed to "assuage domestic or social anxieties" (Acker, 2002; Courtwright, 2001).

As legitimate remedies for a variety of individual and social ills, opioids were constructed as medicine cabinet drugs, rather than street drugs, and were protected by their associations with medicine and the pharmaceutical industry that both manufactured them and widely advertised their curative properties in newspapers. The morphine market was a lucrative one, to the extent that newspapers often refrained from reporting their potentially dangerous effects for fear of losing the precious advertising revenue on which they depended (Duster, 1970). Yet toward the end of the century, some concern began to develop among the medical community and public about the possibility that morphine could be dangerous. At the heart of morphine's danger was the fact that morphinism commonly affected "active brain workers, professional businessmen, teachers, and [other] persons having large responsibilities" (quoted in Conrad & Schneider, 1980) and that it could, as an enemy of professionalism and productivity, pose a threat to the nation. Additionally, another set of concerns circulated around the use of opioids, which involved its transportation to the United States by Chinese immigrants, whose practice of opium smoking was widely disparaged in the popular press. These concerns were undergirded by race, in particular by a widespread anti-Chinese sentiment that accompanied the influx of Chinese immigrants into U.S. cities during these years (Shah, 2001).

In 1898, when the German drug manufacturer Bayer Laboratory began to market another opioid, heroin, sentiments toward opioid use again shifted—in part because of the efforts this major pharmaceutical company and a number of physicians made to celebrate and market the drug as a nonaddictive substitute for opium and morphine. Heroin, unlike its cousin opium, could be efficiently administered with hypodermic needles, which themselves could be readily acquired by anyone who wished to order them out of the Sears catalog. It was also, for many years, available without a prescription (Duster, 1970). Throughout the first years following its introduction, heroin was also lauded as a cure for opioid addiction. Yet this declaration was, within just a few years of its entrance onto the medical market, debunked as more and more physicians began witnessing and reporting on its addictive properties—a trend that also triggered the publication of a host of muckraking exposés, which depicted heroin as "the most threatening drug in history" (Conrad & Schneider, 1980).

By the start of the 20th century, popular sentiments had continued shifting alongside a growing awareness of the emergence of a new demographic of people who use opioids: Early-20th-century heroin users no longer resembled morphine users—those "active brain workers" who were seen as providing the basis of the nation's fortitude and moral integrity. On the contrary, these new users were widely described as "dope fiends," members of a criminal underclass constituted of rough, working-class white males living in urban centers (Valverde, 1998). The emergence of the "dope fiend" in the early 20th century also corresponded with the disassociation of heroin from its legitimate basis in medicine. With heroin use increasingly seen less as a safe medication and more as a recreational drug, public sympathy for people who use opioids declined. Gradually, opioid use became symbolically embedded in the underground world of "the street" and with the danger and moral bankruptcy associated with the illicit drug economy.

The public problem of opioid use, which by then had acquired the meaning of a dangerous vice and a sign of moral weakness, also corresponds to its medical definition during these years. For in medicine, long-term opioid use was also undergoing a transformation in meaning, from the disease of addiction to a kind of psychopathy—a problem inherent to the psychic makeup of the individual person who uses drugs (Kolb, 1927, 1928). As the approach to addiction as psychopathy took center stage, bolstered by the ascendance of psychoanalysis and the psychological study of personality, opioid use was gradually detached from physiological theories and came to be understood as a purely psychological issue. After all, since opioid addiction appeared to have no physical markers, it was seen as an abnormality of individual character—a personality defect that explained the dope fiend's degeneracy, irrationality, criminal nature, and repetitive enactment of certain bad behaviors.

The conceptualization of opioid use as pathological consumption marks a critical moment in the history of opioids for people in pain, one that anticipates the development of a disciplinary approach to opioid use. This approach both regulated opioids and marginalized people in pain who used them as part of a new class of criminal to be "treated" in multifunction prison-hospitals. The passing of the Harrison Act of 1914 is a watershed moment in the disciplinary era of opioid use beginning with the regulation and institutionalization of opioid drugs, which, despite growing concerns regarding their use, were classified as "medications," regulated, and taxed. Six years later in 1920, the beginning of Prohibition institutionalized the criminalization of opioid use with a nationwide ban on all drugs and alcohol. In the same year, moreover, the term "addict" first appeared in the Oxford English Dictionary, as a model of deviance and degeneracy. Even after Prohibition ended in 1933, measures to criminalize drug use and PWUD continued: The Boggs Act, passed in 1941, increased penalties for all crimes involving drugs and established the precedent of minimum mandatory jail sentences for drug offenders—a policy that has only in recent years come under scrutiny in various reform efforts (Administrative Office of the U.S. Courts, 2004, 2007, 2010; Caulkins, 1997; United States Courts, 2017a, 2017b; Families Against Mandatory Minimums, n.d.).

The first two decades of the 20th century inaugurate what David Courtwright (2001) has termed the "Classic Era" of narcotic regulation. In the Classic Era, the "respectable addict" of the 19th century disappeared and was replaced by a vision of people who use opioids as "dope fiends," which was likewise supplemented by the similarly construed figure of the "junkie" (Acker, 2002). For the most part, the mid-20th-century "junkie" resembled the 1920s "dope fiend." Perhaps the only significant differences that exist between the two terms are that, first, the "junkie" applies to a broader demographic of people who use opioids (poor or working-class young men, either white or Black, urban-based, and addicted to heroin) and, second, that this figure not only signifies the psychopathy of addiction but is also meaningful as a representation of creativity and nonconformity within popular culture. As part of the so-called Beat Generation, the "junkie" was not only associated with drugs, but also with a cultural and literary movement launched in the aftermath of World War II. The Beat Generation celebrated, among other things, prolonged experimentation with psychedelic substances. Much of the work that was generated by Beat Generation icons focused on recounting their or others' experiences of drug use and on developing representations of the modern person who uses drugs. Representations of the junkie were, for example, common in the work of Beat writer William Burroughs, in particular his novel *Junkie* (1953), in which the writer reports his interactions with people who use heroin, who he encounters in the inner cities of New York, New Orleans, and Mexico City. His characters are often depraved and, in many cases, deranged. In *Naked Lunch* (1959), Burroughs recalls a day when he made the rounds with a dealer friend of his and describes the depravity coloring the lives of heroin junkies.

> You know how old people lose all shame about eating, and it makes you puke to watch them? Old junkies are the same about junk. They gibber and squeal at the sight of it. The spit hangs off their chin, and their stomach rumbles and all their guts grind in peristalsis while they cook up, dissolving the body's decent skin, you expect any moment a great blob of protoplasm will flop right out and surround the junk. Really disgust you to see it (p. 16).

As a self-declared heroin consumer, Burroughs' description of the wantonness of the typical junkie is consistent in his recounting of his own experience with opioid use, where he describes how his heroin habit transformed him into a "lush worker," a low-life degenerate who spent most of his time in the criminal underworld envisioned through the New York City subway system, where he learned to identify passengers who were drunk, passed out, or sleeping, and casually steal their wallets.

The image of people who use heroin as degenerate "junkies" circulated not only in popular parlance but also in medical research and policy (Acker, 2002; Campbell, 2007). Drug historian Nancy Campbell (2007) has written that during this period, addiction research was not an academic exercise but a governmental one, taking place within federal institutions like the Bureau of Social Hygiene and the Bureau of Narcotics.

The deployment of a psychojuridical or disciplinary model of addiction is likewise illustrated in the government's funding and construction of its infamous "narcotic farms" in Lexington, Kentucky, and Fort Worth, Texas. Jointly operated by the Public Health Service and the Federal Bureau of Prisons, these "farms" were essentially prison-type institutions masquerading as hospitals (Campbell, 2007). As Campbell explains, the narcotic farms were also the privileged sites for the disciplinary normalization of opioid use and addiction, where treatment was carried out through an individualizing and punishing mode of intervention in which blame was focused on the internal shortcomings of the person who uses opioids themselves. And while the disciplinary institutions within which people who use opioids were confined and "treated" were not exactly prisons, they were funded by the federal government and (in the words of one key psychiatrist who worked there) "operated like a hospital" that "necessarily had prison features" (Acker, 2002). As part-prisons, part-hospitals, the narcotic farms stand in as an early example of the blurring of the lines used to separate patients from addicts, disease from addiction, and treatment from punishment.

The internment of people who use opioids in narcotic farms predates but, in some ways, anticipates the start of the U.S. government's "war on drugs" campaign. The war, which was declared by Nixon in 1970, carried forth by President Reagan, and waged by nearly every other president since, has been primarily characterized by the widespread criminalization of drug use and a new, "tough" stance toward the illegal drug trade. And it has been waged in particular against drug sellers and Black PWUD (Alexander, 2010/2012; Marez, 2004).

The explicitly punitive approach to addiction in the 1960s, '70s, and '80s is, in some ways, a curious one, considering that during these same years the psychological approach to drug use that likened the PWUD's personality to psychopathy was supplemented and largely replaced by a concept of drug use that was biologically and physiologically based. The first of these approaches, a metabolic theory of addiction, developed by Vincent Dole and Marie Nyswander in 1967, became a dominant approach to addiction in both medicine and psychiatry. Dole and Nyswander's theory essentially said that addiction is not caused by an addictive personality but by a neurological susceptibility and an altered response to narcotics. As proponents of a neurological model of addiction, Dole and Nyswander denounced the "moralist" approach to it and advocated instead for pharmacologically based research and treatment. For the past 20 years, addiction research that is neurobiologically oriented—led by the National Institute on Drug Abuse (NIDA)—has promoted an understanding of addiction as a "chronic, relapsing brain disease." In accordance with this model, drug use is understood as the "behavioral outcome of a biological dysfunction," one in which the brain's systems for reward, motivation, learning, and choice do not function as they should (Raikhel & Garriott, 2013, p. 13). Hence, drug addiction manifests itself in a series of behaviors that include the "compulsive seeking... and administration of a drug despite grave adverse consequences" (Nestler, 2004, p. 698). This model is today bolstered by the recent development of neuroimaging technologies, tools that work to trace drug use back to dysfunctional brain systems, which can be shown to "light up" when an individual consumes an addictive substance or engages in an addictive behavior.

The neuroscientific approach to drug use and addiction has been characterized by many as an attempt to demolish the moral stigma and social injustices associated with past approaches, particularly those that have explicitly targeted and led to the widespread incarceration of Black and Latino(a) PWUD. Yet while it may be the case that the current conceptualization of addiction is based in neuroscientific research, which attempts to define it in biological terms that disassociate it from psychopathy and criminality, it is also the case that moralizing approaches to drug use have not disappeared. And biology, which undergirds the neuroscientific approach to addiction, has not necessarily succeeded in disassociating drug use from its social, moral, or criminal connotations. On the contrary, as many have argued (e.g., Campbell, 2010; Dumit, 2004; Netherland & Hansen, 2016; Vrecko, 2010), addiction neuroscience has also played a role in extending the social injustices and racial stratification associated with drug use.

Discourses that criminalize opioid use, which associate it with problems in an individual's character or with inherent moral weakness, still circulate in popular culture, politics, and the law. In these domains, however, such discourses now exist in tension alongside neurological models that situate opioid use and addiction within the brain of the person who uses opioids. Yet though these two approaches may seem opposed to one another, they do in fact agree on a pair of key points: First, both share a concern with behaviors, especially those that, because they are linked to compulsive self-harm, are seen as violating the terms of human choice, free will, and self-control. Second, both understand addiction to be something internal, which is situated inside the PWUD—whether in their brain or their character.

Whether they belong to the "at-risk" / "consumer" / "consumed" typology I have outlined here or whether they are made to fall along the more dichotomous lines of chronic pain "patients" or opioid "addicts," all people in pain (especially those who use opioids) are subject to definition. Moreover, the definitions that are attached to them—by governments, lawmakers, medical practitioners, and sometimes, their fellow citizens—have stunning material consequences. We see these consequences take shape in public health, where the uneven distribution of life-saving resources for people who use opioids is marked by race and class differences. We see them play out in politics, where huge differences in lobbying power favor the needs of chronic-pain patient advocacy groups over those of people who use illicit opioids. We see how they redefine the power dynamics animating the failed "war on drugs," which continue to shift alongside recent trends in illicit fentanyl and heroin use. Of course, these are not the only areas in which the discursive construction of people in pain have borne out material consequences, yet they are particularly salient in the opioid epidemic. And they provide illustrative answers to this chapter's most pressing question: Whose pain is made to matter, and whose is not?

Whose Pain Matters?

In history, politics, and popular culture, attempts to distinguish legitimate suffering from illegitimate pain abound. We see them emerge from discussions around pain undertreatment, opioid addiction, and welfare. For chronic pain patients, just as for people in pain who use drugs outside of the medical system, gaining legitimacy has long been a problem. Only one of these populations, however, has been successful in gaining recognition, resources, and respect. The power imbalance that splits chronic pain patients from other kinds of people who use opioids is thus worth taking time to consider, as it provides a case study that illustrates some of the key power dynamics that have been central to the opioid crisis and that continue to characterize the world of opioid use, as overdose death rates continue to rise.

The power imbalance between chronic pain sufferers who have been diagnosed defined as patients versus those who use opioids but have not been diagnosed, lies in part in the former group's associations with big pharma. According to a U.S. Senate report, 14 patient advocacy organizations and health professional organizations for people in chronic pain received

$9 million dollars from 2012 to 2017 from pharmaceutical donors (HSGAC, 2018). Purdue Pharma was, notably, the largest donor, giving more than $4 million to these groups. Among them was the U.S. Pain Foundation, which received nearly $3 million, the Academy of Integrative Pain Management, which received upward of $1 million, and the American Pain Society, which earned nearly $1 million from opioid companies. As Marks (2020) has written, another one of these organizations, the American Pain Foundation, has continued to accept donations from the opioid industry even after closing its doors in 2012 in the wake of investigations into its dependence on opioid manufacturer funding. Although their ties to opioid manufacturers can render them vulnerable to outside criticism, many chronic pain patient advocacy groups have benefited from forming alliances with the opioid industry, whose $880 million lobbying spending between 2006 and 2015 to fight against opioid prescribing limits completely dwarfed the $4 million spent by groups advocating on behalf of those limits (CPI, 2016).

The lobbying power of chronic pain patient organizations has enabled these groups to successfully influence opioid prescribing rules and drug policies, including those set forth by the CDC. For example, in 2016, the CDC issued a set of new guidelines for opioid prescribers, which attempted to restrict and limit the number of opioid prescriptions. Chronic pain patient groups were understandably outraged by the new limits set on opioid prescribing practices and immediately began organizing to appeal them. After three years of extensive lobbying, the chronic pain community convinced the CDC to act to revise its guidelines. In its revisions, the CDC explicitly stated that chronic pain patients are exempt from opioid restrictions, and that its guidelines were never intended for people who are suffering from acute and chronic pain. In April 2019, chronic pain advocacy groups gained another win, successfully convincing the FDA to change its labeling of opioid products to alert clinicians that curtailing or tapering opioid prescriptions for long-term chronic pain patients is not only ill-advised but also dangerous and can result in "serious withdrawal symptoms, uncontrolled pain, psychological distress, and suicide" (Wallis, 2019).

It is worth noting that, in the CDC guidelines and FDA labeling changes, nothing is said about people who use illicit opioids and how the guidelines should be applied to them. In fact, this population is auspiciously absent from official debates about pain undertreatment and opioid prescribing. In large part, this has to do with the fact that PWUD tend to lack the financial leverage and ties to the pharmaceutical industry and professional organizations that have often empowered chronic-pain patient organizations and helped them to successfully push pro-opioid policies.

Indeed, for many years, debates about chronic pain undertreatment have focused rather myopically on the suffering of legitimate chronic pain patients, rather than on "illegitimate" PWUD, as the population most in need of pro-opioid drug policies and reforms. Many of the notable pain management experts mentioned in this book's previous chapters have been involved in these discussions and have demonstrated a strong commitment to the goal of reforming state laws, regulations, and medical board policies regarding opioids (Herzberg, 2020). Two particularly vocal advocates, David Joranson and David Haddox, both pushed—in parallel with opioid manufacturers—to challenge the restrictive opioid consensus that had dominated the medical field for decades. Both men have played key roles in the creation and growth of chronic pain patient groups and professional organizations such as the American Association of Pain Management and the American Pain Society, which joined together in 1997 to publish a consensus statement challenging restrictive opioid prescribing practices. Haddox, who, as previously mentioned, invented the concept of "pseudo-addiction," chaired the committee that wrote the statement, which debunked the long-held consensus that opioids are addictive and pushed for the easing of guidelines to make opioid medications more readily available to chronic pain patients. The co-chair of the committee, pain expert David Joranson, was likewise involved in the fight to ease opioid restrictions and eventually organized the Pain & Policy Study Group (PPSG) at the University of Wisconsin-Madison to research and mount challenges against drug policies that hindered chronic pain patients' access to opioids. The PPSG received millions of dollars in funding from pharmaceutical companies, including Purdue Pharma (Herzberg, 2020).

Thus, while chronic pain advocacy can be demonstrated to have leveraged considerable financial resources and political power in its fight against antiopioid drug policies, PWUD have been largely invisible in these policy battles. Their invisibility is linked, in part, to the fact that while pharmaceutical companies and patient/professional organizations were lobbying to liberate opioids for one population (chronic pain patients), so too were they engaged in a long-standing attempt to distance themselves from another (PWUD). In fact, the problem of PWUD and, in particular, the association of this group with opioid addiction, has long haunted pain management. Since at least the 1980s, the field has been involved in a concerted effort to distance long-term opioid use and opioid tolerance from the stigma of addiction.

Across much of the literature in pain management, opioid addiction has been central to the problematization of pain—particularly as it has related to the concerns held by pain researchers and physicians regarding the widespread undertreatment of cancer and non-cancer-related chronic pain (e.g., Dahl, 1997; Foley, 1985; Gilson & Joranson, 20021; Halpern & Robinson, 1985; Jaffe, 1980, 1989; Joranson & Gilson, 1994; Joranson et al., 2002; Portenoy, 1996; Portenoy & Foley, 1986). For example, in a paper published in the *Journal of Pain and Symptom Management* in 1996, eminent pain specialist Russel Portenoy included the conceptualization of opioid addiction as one of the "critical issues" in contemporary pain management. Portenoy argued that coming to grips with the prevalence of chronic pain undertreatment would require pain specialists to recognize that the "confusing nomenclature" used to define opioid addiction is a hindrance to successful pain management (p. 208). Undertreatment, Portenoy insisted, results in part from the widely held (and erroneous) belief held by most clinicians and physicians that an individual whose physical tolerance to opioids has increased is, by virtue of that tolerance, addicted to these drugs. The conflation of physical tolerance and addiction, as Portenoy saw it, is what undergirds many physicians' fears about people who use opioids and subsequently leads them to refuse their patients' requests for adequate care and pain relief. To reverse the cycle of undertreatment, he and others (e.g., Dahl, 1997; Foley, 1985) advocated for those working in pain management to engage in a sustained attempt to redefine the concept of opioid addiction in order "to place the use of narcotic analgesics in perspective" and put to bed, once and for all, the stigma associated with people who use opioids (Portenoy, 1996, p. 88).

The critique that was launched within pain medicine, which argued that defining opioid addiction in terms of opioid tolerance (a nearly inevitable result of long-term opioid use) was exacerbating the problem of undertreatment for many chronic pain patients, also coincided with efforts to redefine and reframe opioid addiction in the interrelated domains of the law, policy, and regulation (Gilson & Joranson, 2002; Joranson & Gilson, 1994; Joranson et al., 2002). Discussions of opioid addiction in these domains, as Joranson and others have argued, often failed to define an addicted condition—which would set a precedent for its treatment—and instead was fixated on conceptualizing addiction as a particular kind of person in pain. For example, the 1988 Modern Medical Practice Act (MMPA), published by Federation of State Medical Boards (FSMB), referred to "an habitué or addict or any person previously drug dependent any drug legally classified as a controlled substance or recognized as an addictive or dangerous drug" (quoted on p. 218). As many pain specialists and advocates have suggested, such definitions have a performative effect—one that directly affects the lived experience of chronic pain patients and makes it more difficult for them to get the drugs they want or need without being severely policed, regulated, and monitored.

While the above critiques have certainly helped many people who need opioids gain access to them, and while they may have diminished the stigma of opioid addiction among chronic pain patients, these efforts have largely sidelined a large population of PWUD who do not have a diagnosis or prescription. These are people who, in line with the figure of the pathological consumer described at the beginning of this chapter, tend to fall outside the sanitized realm of the opioid "white market" (Herzberg, 2020). They are the people who lack ties to rich pharmaceutical companies, to renowned medical experts, and to advocacy organizations swimming in resources. For these well-positioned and well-resourced entities, the push to acknowledge the problem (and it *is* a problem) of pain undertreatment has been largely successful. Prescribing limits have been loosened, and chronic pain is, today, recognized as a legitimate medical problem warranting a wide array of solutions. Yet the push to confront pain undertreatment, insofar as it concerns only those who are classified as patients, renders the suffering of other people in pain, those who use drugs but who are not considered to be patients, as undeserving of our attention.

The Case of Naloxone

That the pain of PWUD has, before and throughout the opioid epidemic, been seen as unworthy of attention and advocacy is perhaps most evident in the story of naloxone, an opioid antagonist medication used to reverse the effects of opioid overdose. First approved by the FDA in 1971, naloxone did not make its way into public health and harm reduction drug policy until more than three decades later. Nancy Campbell (2019) has written extensively about naloxone, and about the delayed recognition of its potential as a harm reduction measure. For Campbell, the slow realization that naloxone could be used to save the lives of PWUD does not point to a success story about drug discovery but, rather, to decades of struggle over what constitutes proper drug use, including when naloxone should be used, who can administer it, and who deserves to receive it. The eventual success of naloxone is attributed, moreover, not to the kinds of alliances we saw taking shape among patient advocacy groups, professional organizations, and opioid manufacturers, but, rather, through the grassroots efforts of drug activists, who rely on "fragile alliances between social movements, scientists, and government actors" (Campbell, 2019).

In 2012, activists finally gained ground in the debates about opioid use when the FDA held its first hearing on possible regulatory approval for naloxone as an over-the-counter medication. Even then, however, activists had still not achieved enough power or visibility to be invited to sit at the regulators' table (Campbell, 2019). A year later, the Substance Abuse and Mental Health Services Administration (SAMHSA) developed an overdose prevention kit that advised clinicians to coprescribe naloxone to patients taking opioids, regardless of whether the patients were receiving long-term or high-dose opioid therapy. The effectiveness of naloxone and the critical need for it during the opioid crisis also prompted the U.S. Surgeon General to issue a public health advisory in April 2018, which recommended increased availability of the medication, especially in communities with high rates of opioid use. The advisory also recommended administration by a wide range of health officials, first responders, PWUDs, and their family members, a significant step in increasing and broadening access (HHS, 2018).

Even while naloxone became more widely used and prescribed in the 2010s, there was significant debate over whether the medication would augment problematic drug use among people who use opioids. One article published in 2018 argued that it almost certainly would. Doleac and Mukherjee's article, "The Effects of Naloxone Access Laws on Opioid Abuse, Mortality, and Crime," argued that naloxone access unintentionally increases opioid abuse, primarily through two channels: First, it does so by reducing the risk of opioid use, "thereby making riskier opioid use more appealing." Correspondingly, the authors argue that naloxone increases the incidence of crime, particularly theft, as it increases "the number of opioid users who need to fund their drug purchases." But perhaps most shocking is authors' assertion that the second channel through which opioid misuse is increased, which they identify as the most problematic, is the lives that are saved by it. In other words, it is by "saving the lives of active drug users" that naloxone exerts its "negative net effect."

Doleac and Mukherjee's article, which has since been critiqued for its indiscriminate use of indirect data (including Google searches for "naloxone" and "drug rehab"), offers a clear example of the biopolitical thrust that has long driven the war on drugs, and that continues to be used in drug policies that discriminate against and seek to punish PWUD. In this biopolitical regime, the lives of PWUDs are not seen as inherently worthy of saving but, rather, as a potentially "negative net effect," one that should make policymakers think twice about the impact of the harm reduction policies they intend to implement. The 2018 article is likewise illustrative of what Jerrett Zigon (2018) has argued in his book about the anti–drug war political activity and the logics undergirding the failed war on drugs—that drug wars are, ultimately, a war on certain kinds of people. In this case, it is PWUD whose pain, whose lives, and whose inherent worthiness of being saved are called up for debate.

While naloxone has, thanks to grassroots organizing by drug activists, become increasingly available and accessible to PWUD, it is also the case that the price of this essential harm reduction tool has increased astronomically since its approval by the FDA. From 2011 to 2018, naloxone manufacturer Hospira increased the price of naloxone by over 1,000%. And in 2013, the same year that SAMHSA issued guidelines advising the over-the-counter prescription of the medication, Hospira again raised the price of naloxone to be purchased by the New York State Department of Health by an additional 43%. While in 2006 a single dose cost the state of New York just over a dollar, it would now cost $13.50, a price hike that drastically reduced the medication's availability both in New York City and the rest of the state (Zigon, 2018; Gupta et al., 2016). Newer, more easily administered formulations of naloxone are even more expensive. Narcan, as Zigon writes, costs $150 for just two nasal spray doses. And a two-dose package of Evzio cost $690 in 2014 but is now more than $4,500, a price increase of about fivefold.

Moreover, while naloxone distribution has increased over the past five years, even doubling between 2017 and 2018, wide variations in dispensing in pharmacies exist, despite consistent state laws. In fact, naloxone dispensing is 25 times greater in the highest-dispensing counties than in the lowest-dispensing (CDC, 2019). In a 2019 press release, the CDC noted that in the year prior, rural counties across the country had the lowest naloxone dispensing rates, with providers writing just 1.5 naloxone prescriptions for every 100 high-dose opioid prescriptions, over half of which required a co-pay. In comparison, metropolitan counties were three times less likely to be a low-dispensing county. Nationwide, only one naloxone prescription is dispensed for every 70 high-dose opioid prescriptions. And nearly 70% of Medicare prescriptions for naloxone require a co-pay, compared with

42% for commercial insurance (CDC, 2019). There are also racial and class discrepancies in naloxone access. One study based in New York City found that the residential area with the highest proportion of Black and Latinx low-income individuals also had the highest methadone treatment rate. In comparison, buprenorphine and naloxone were most accessible in residential areas with the greatest proportion of white, high-income patients (Hanson et al., 2013).

Fentanylization and a Return to the War on Drugs

Alongside naloxone, story of fentanyl also illustrates how, in the opioid crisis, some people's pain is rendered worthy of attention, while others' is not. Of the nearly 70,000 opioid-related overdose deaths in 2020, nearly 57,000 were attributed to fentanyl (NIDA, 2022). Indeed, the landscape of opioid use has shifted drastically since the early 2010s, and PWUDs have increasingly been moving away from pharmaceutical opioids toward more readily available and less expensive heroin and synthetic opioids like fentanyl and its analogs. The fentanylization of opioid use has eroded much of the sympathy that was initially attached to prescription opioid victims, pointing to the ways in which power dynamics and shifting cultural narratives condition the ways in which a person's pain will be recognized as legitimate and worthy of treating—or not.

Before fentanyl took over the opioid landscape around 2011–2013, lawmakers and policymakers seemed to be indicating a shift in their approach to drug policy. These years were marked by a rhetorical emphasis on a "gentler war on drugs" and a "compassionate approach to drug use" that advocates for treatment over imprisonment (Seelye, 2015). However, since fentanyl's entrance into the U.S. opioid market, the actions that law- and policymakers have taken to combat the epidemic have become increasingly punitive, advocating for draconian policies that persisted—and are now being expanded—throughout Nixon, Reagan, and Clinton's botched war on drugs.

Since 2011, 45 states have proposed legislation to increase penalties for fentanyl. Thirty-nine states and Washington, DC, have passed or enacted such legislation. Similar attempts to crack down on illicit fentanyl use have been launched at the federal level: For example, in March 2017, the President's Commission on Combating Drug Addiction recommended increasing penalties for trafficking fentanyl and its analogs. And one year later, President Trump signed into law new legislation granting new detection tools for the Department of Homeland Security, in order to identify and seize fentanyl at the U.S. borders, referring to the drug as "our new big scourge." Trump also indicated in a speech he gave in New Hampshire that he would support the death penalty for people who sell illicit drugs (Merica, 2018).

The Trump administration's crackdown on fentanyl and PWUD has also been driven by efforts advanced by former Attorney General Jeff Sessions, who often leveraged fentanyl to justify his escalation of policies reminiscent of the war on drugs and undo Obama-era sentencing reforms (Collins & Vakharia, 2020). In the summer of 2017, Sessions proposed that the DEA mobilize its own team of prosecutors to focus exclusively on opioid cases. While the DEA has never prosecuted crimes, Sessions' proposal would grant the agency the power to do so for federal drug cases. The move was strongly opposed by Democrats in the Senate, however, with New Jersey Senator Corey Booker referring to the move as "a thinly veiled attempt to ramp up [the] failed War on Drugs" (Owen, 2017). In 2018, Sessions announced the launch of the so-called Operation Synthetic Opioid Surge (SOS), which instituted an "enforcement surge" in 10 different areas of the United States that grants powers to attorneys general to begin prosecuting every "readily provable case involving the distribution of fentanyl, fentanyl analogues, and other synthetic opioids, regardless of drug quantity" (U.S. Department of Justice, 2018). Since Session's resignation, his successor William Barr has maintained his punitive stance toward opioid use, prioritizing the prosecution of people who buy and sell these substances. According to a report by the Drug Policy Alliance, the Trump administration oversaw a 40-fold increase in federal fentanyl-related prosecutions during the president's short time in the White House (Collins & Vakharia, 2020).

Despite this recent trend toward drug war–era criminalization, it is important to note that the harm reduction movement, which is opposed to punitive drug policies, has gained some ground in recent years, during which it has experienced considerable success in passing laws and implementing interventions to reduce the dangers that accompany long-term drug use. Yet at the same time, the past several years have revealed a consistent backtracking of efforts to reform the criminal justice system, much of which has been inspired by the recent fentanylization of opioid use. The Trump administration's tough stance toward opioids is also aligned with a broader public sentiment, in which sympathy for certain kinds of drug users—particularly those who use fentanyl and heroin—has begun to erode. Media accounts of fentanyl that depict it as a dangerous poison and the people who use it as equally dangerous traffickers help to fuel popular support for the government's punitive approach toward people who use opioids. This support is also augmented by waves of fentanyl-related misinformation that have circulated widely in the press.

One such falsehood, which was initially reported by the San Diego County Sheriff's Department and later spread throughout social media and other news outlets such as the *New York Times*, featured a video that showed a police officer collapsing and "nearly dying" after coming into skin contact with a fentanyl analog (Wolford, 2021). However, according to medical experts, overdosing after merely touching fentanyl is essentially impossible. And yet the story blew up. The 4-minute video of the officer's "overdose" spread rapidly throughout the Internet, spawning a moral panic about the dangers of fentanyl and the people who consume and distribute it (Beletsky et al., 2020). Fentanyl's hypervisibility in the media, combined with the proliferation of false narratives about it, serve to rationalize punitive drug policies and curtail access to a drug that many people—regardless of their prescription status—still need.

* * *

So, whose pain is it that really matters? Whose doesn't? As this chapter has tried to show, one cannot begin to understand the opioid epidemic and the way in which it has been constructed in medicine, politics, and popular culture without also attempting to answer these questions. Doing so, moreover, reveals a long history of attempts to delineate and distinguish the pain of certain kinds of people who use opioids from others. These attempts to define people who use opioids as different kinds of consumers have, for many decades, functioned in ways that have ended up determining which kinds of people who use opioids are worthy—of medical treatment, of compassion, or of mere recognition. Time and time again, the answer to this question has revolved around patienthood and on identifying those individuals whose opioid use adheres to a particular set of social, legal, and medical norms. But what about the rest of the people in pain? Whether they use OxyContin or fentanyl or heroin, shouldn't their pain matter too? There is an obvious answer to this question, but as this chapter has tried to demonstrate, history, policy, and practice seem to suggest otherwise. In the stark power imbalance that animates the relationship between patient advocacy groups and PWUD, as well as the debates around naloxone and the drug policies that correspond to the fentanylization of opioid use, too many people in pain continue to suffer. Oftentimes, they go at it alone—unrecognized, not advocated for, and, frequently, castigated for attempting to attend to what is perhaps the most basic human need, to avoid pain.

Closing

I began this research with the intent to examine and understand what has come to be known as the "opioid epidemic." I wanted to identify the key cultural and institutional conditions that contributed to its formation and assess how well these conditions helped to explain the incredible upsurges we have seen in the manufacture, promotion, prescription, consumption, and overdoses related to opioids in the United States over the last 20-plus years. And I wanted to know how the use of opioids in each of these contexts was legitimized and maintained, particularly amid growing public discourse about the risks associated with them. As I dug deeper into these questions, it became clear to me that to answer them I would have to think more carefully about the problem of pain—a problem that, as this book has argued, lies at the heart of the opioid epidemic. Pain is, moreover, a crucial starting point for understanding the place that this crisis has come to occupy in American culture.

The question of how the opioid epidemic relates to broader social and cultural issues is crucial since it illustrates the significance of what might otherwise be understood as a predominantly medical phenomenon. Developing a deeper and broader understanding of the opioid epidemic and—more specifically, of the problem of pain—has constituted one of this project's main thrusts. Throughout these pages, I have sought to provide an alternative lens to examine what has undoubtedly become one of the most formidable social issues of our time. To analyze it from the perspective of medicine or science alone risks oversimplifying the complexity and subtlety of its underlying dynamics. And while I don't mean to suggest that arguments rooted in biology and physiology can't enrich our understanding of pain, it does seem to me that such arguments become even more useful once they are considered in light of the role that culture plays in structuring scientific and medical truths. This is largely because medical and scientific truths are anything but fixed. On the contrary, they are in flux and can be shown to shift over time and in connection with other social, cultural, and political developments. Thus, opening medical and scientific categories to critical examination is essential if we are to understand how they become "common sense," how they structure human lives, and the roles they have come to play in the reproduction of norms, values, and hierarchies within our society.

But the reproduction of norms, values, and hierarchies not only works directly through the production of scientific/medical knowledge; it is also a cultural process, one that is driven by communication, representation, and experience. Developing this project has meant attending to these processes—for at every turn, pain, addiction, and opioids have become entangled with new meanings and associations that dramatically altered their functions as well as their legibility. As a cultural phenomenon, the opioid epidemic is a site where power relations have been and continue to be constantly exercised and negotiated. The ways in which power relations are mobilized in and through the opioid epidemic are perhaps clearest when we consider the ways in which the construction of different kinds of people who use opioids reproduces inequalities along the lines of race and class.

Thus, at the heart of this project lies a critical interrogation of health, science, and medicine. This is in part because examining the culture and politics of pain gives us cause to question (and perhaps even to reject) the idea that scientific and medical knowledge about it is unbiased or neutral and that it is somehow more valuable than other kinds of knowledge. For on the contrary, expert knowledge about pain in these domains is laden with biases and assumptions about which kinds of suffering matter, for whom they matter, and who gets to decide.

Rerouting Addiction Through the Problem of Pain

The problematization of pain is connected to what I believe to be a significant shift in cultural and political discourse related to addiction. The problem of pain is driving a shift in the rhetoric and practices related to addiction and people who used drugs (PWUD). In particular, the locus of blame for addiction is becoming increasingly detached from the supposed internal flaws of addicts themselves and repositioned onto a variety of actors, institutions, and other forces external to them, within the country's medical–industrial complex. While scrutiny of the medical and pharmaceutical industries is not anything new, it is taken to an entirely new level in the opioid epidemic. This is reflected in a rather stunning series of recent events that are pushing drug companies to acknowledge the role they have played in the current crisis. While, typically, legal battles waged against Big Pharma don't result in a significant change in company practices, recent lawsuits that have forced some opioid manufacturers to stop advertising their top-selling drugs and have pushed others into bankruptcy speak volumes about the indignation and outrage that continues to swell among a growing American public that is increasingly affected by the products and actions of these companies. These events mark the start of what looks to be a concerted attempt to intervene directly into the opioid market. As is discussed in this book's analysis of opioid regulation, such an interventionist tactic runs counter to the logic of laissez-faire regulation and the role it has played within the opioid crisis and United States' healthcare system as a whole.

As previously stated, the growing prevalence of opioid addiction and the profound connection it shares with the problem of pain are articulated to the broader landscape of American culture, where these problems are shaped by (and in turn shape) key dynamics within this wider field. These include, but are not limited to, mounting concerns and attention related to the deterioration of America's white working class. The growing rates at which this demographic has been ravaged by so-called deaths of despair—a phenomenon that plays an important role in making the opioid epidemic legible as a national emergency—also says something about what Raymond Williams (1977) referred to as the "structures of feeling" of our current moment. Collective concerns and pessimism about the future crystallize in the opioid crisis, where pain becomes visible not so much as a medical fact but as a powerful affect, one that has much to say about how many Americans feel about themselves, others, and their relationship to a tumultuous world.

It seems important (though perhaps obvious) to note that the centering of pain as a key affect within today's structures of feeling must be viewed ambivalently, since the experience of pain is something that has the ability both to draw connections among people and to divide them. As such, discourses formed around the problem of pain both reflect, maintain, and, in some cases, perpetuate social, cultural, and political rifts. At the same time, the experience of pain is perhaps the only experience available to all people and, as such, could provide a productive starting point for addressing social inequalities and injustices—beginning with those that are reinforced within the U.S. healthcare system.

Drug Policy and Implications for the Future

To address the growing opioid epidemic, policymakers have focused largely on controlling the prescription and use of opioid analgesics through the implementation of supply-side drug policies and attempts to limit opioid prescribing and reduce access to prescription opioids. Thirty-eight states have opioid prescription limit laws, which impose restrictions on initial opioid prescriptions. These policies often limit initial prescriptions to seven days and set a cap on the dose that can be prescribed. Another common supply-side drug policy is pain clinic laws, attempts to crack down on "pill mills," which include the registration of pain clinics with the state, requiring physician ownership of the clinics, prescribing restrictions, and record-keeping requirements. Additionally, states have implemented prescription drug monitoring programs (PDMPs) as a solution to addressing the opioid crisis. Yet even though today all 50 states have PDMPs in place, the national death rate due to opioid overdoses has not decreased. This is in part because PDMPs are meant to catch people who are "doctor shopping," and just a small percentage of people who use opioids obtain them in this way. As discussed in Chapters 3 and 4 of this book, most people who use opioids for the first time either receive them from one doctor or from illicit and semi-illicit channels. In their systematic review of studies evaluating the effectiveness of PDMPs, Fink and colleagues showed that half the studies examining the relationship between PDMPs and heroin overdose found nonsignificant effects, while the other half found that PDMP implementation was associated with increases in heroin overdoses (Fink et al., 2018). Finally, regulators have also intervened in drug policy by reformulating prescription opioids as abuse-deterrent medications, which, as Chapter 2 detailed, attempted to limit opioid use but were largely unsuccessful. Instead of discouraging opioid use, the introduction of abuse-deterrent (opioid) formulations (ADFs) had a boomerang or balloon effect, pushing people who use opioids toward illicit, unregulated substances like heroin and fentanyl.

In tandem with policies that attempt to reduce opioid use, some policy measures have recently been enacted that focus on reducing harms associated with treating and reporting drug overdoses. One such strategy, naloxone access laws, allow for the prescribing of naloxone to individuals with a documented substance use disorder, allow for the dispensing of naloxone at pharmacies, and provide immunity to certain individuals who dispense and administer naloxone. Harm reduction policies also include Good Samaritan laws, which provide immunity to people who call 911 or seek emergency medical assistance for an overdose. All 50 states have enacted some type of Good Samaritan law in lieu of the opioid overdose crisis (West & Varacallo, 2021).

Public opinion has also been shifting with regard to drug policy. According to a 2014 Pew Research survey, 67% of Americans said that the government should focus more on providing treatment for those who use illicit drugs such as heroin and cocaine. Just 26% thought the government's focus should be on prosecuting people who use these drugs (Pew Research Center, 2014a). Yet despite this shift in public sentiment about drug use, drug policy has continued to focus on suppressing access to opioids. To that end, the criminal justice system has stepped in and intensified its focus on arresting, prosecuting, and incarcerating people who use and sell illicit drugs (Beletsky & Davis, 2017; Davis, 2017). For example, President Trump and his administration instituted several policies that further criminalize drug use, especially as it involves illicit opioids. Based on statements from Trump and members of his administration, the former president threatened to reinstate many of 1980s-style drug war policies, including harsher sentences for those who sell and consume illicit drugs. Trump also called for the death penalty for people who sell drugs and, famously, started building a wall on the premise of keeping drugs out of the United States. Alongside its Reaganesque "just say no" messaging, which was especially targeted at young people, the Trump administration enacted legislation to apply a class-wide emergency scheduling of fentanyl-related substances in the Stopping Overdoses of Fentanyl Analogues Act of 2019 (SOFA) and the FIGHT Act (Collins & Vakharia, 2020).

In contrast, the Biden administration has stated that it does not support Drug War–era policies that focus on incarceration and the criminalization of PWUD, and President Biden specifically opposed a 1994 crime bill intended to punish PWUD when he was in the U.S. Senate. The Biden administration has advocated for a "compassionate approach" to drug use that focuses on "ensuring racial equity in drug policy and promoting harm-reduction efforts" (ONDCP, 2021). Yet at the same time, the Biden administration has also doubled down on fentanyl and the people who use and sell it. In May 2021, President Biden signed into law Trump's failed "Extending Temporary Emergency Scheduling of Fentanyl Analogues Act," which allowed for a temporary scheduling order classifying certain fentanyl-related substances as Schedule I drugs subject to the strictest controls. Over 140 groups and organizations wrote a letter imploring the president to let the act expire in October 2021 (Human Rights Watch, 2021). Yet since then, Biden has urged Congress to make Trump's policy permanent, a decision that would further criminalize many people who use opioids.

Even while policymakers' discourse and public sentiment have shifted over the past 10 years toward a less punitive model of drug policy, much of the original infrastructure from the war on drugs has been maintained. And many of the policies that criminalize drug use are being used to target people who use or sell fentanyl and heroin. In particular, the rise of fentanyl has paralleled an uptick in support for mandatory minimum sentences, drug-induced homicide laws, and excessive police budgets that go together with militarized law enforcement, which is largely funded by the Department of Defense's "1033 Program" and the Department of Justice's Edward Byrne Memorial Justice Assistance Grant (JAG). Alongside these strategies are other punitive measures, including the excessive use of surveillance by law enforcement, which leads to unreasonable search and seizure operations and home invasions like that which led to the death of Breonna Taylor in March 2020. While policymakers in the United States often recognize that drug use should be treated as a public health issue, many continue to support harsh criminal sentences for people who sell or distribute drugs. Crackdowns on people who sell drugs are advanced under the pretense that these individuals are causing the overdose crisis. However, it is demand, not supply, that drives the drug market (Drug Policy Alliance, 2019). Additionally, much drug policy defines "drug dealers" broadly and neglects to consider the overlap between PWUD and people who sell or distribute drugs. In fact, according to one 2012 survey, 43% of people who sell drugs also use them and meet the criteria for substance use disorder (Stanforth et al., 2016).

These Drug War policies have extremely negative effects. By ignoring the overlap between people who use drugs and people who sell them, policies that opt for increased penalties for "drug dealers" end up criminalizing PWUD. Criminalizing supply-side drug market activity, moreover, can make drug use more dangerous and can contribute to overdoses by incentivizing the development and consumption of more potent, riskier drugs like fentanyl and its analogs. Such policies also undermine Good Samaritan laws and discourage people from calling 911 when they witness an overdose. Drug prohibition also increases the possibility for violence, in part because it drastically increases the value of drugs and incentivizes the entrance of large criminal organizations into the drug market (Jacques & Allen, 2015; Werb et al., 2011). Finally, harsher drug sentencing and excessive permissions for law enforcement are disproportionately used against people in poverty, people of color, noncitizens, and women (Drug Policy Alliance, 2019). In 2019, among people who were sentenced for fentanyl-related offenses, 40.5% were Black, while 24.3% were white, despite the fact that racial groups buy and sell illicit substances at similar rates (United States Sentencing Commission, 2021; Hamilton Project, 2016).

The criminalization of drug use and PWUD impacts every sector of our lives, from housing and education to immigration, employment, and child welfare. The Drug Policy Alliance (2021) recently published a report that outlines some of the concrete policies that negatively impact the lives of PWUD: First, in 18 states employers are allowed to conduct drug tests on their employees, regardless of that employee's job function, which often leads to the termination of PWUD and the maintenance of cycles of poverty and inequality. Moreover, half of all states, including the District of Columbia, require doctors to report suspicions of drug use to child welfare authorities, which puts parents and their children at risk of being separated. In the education sector, more than 10 million students go to schools that have law enforcement officers but no social workers, a fact that reflects the ways in which drug policy is still prioritizing the aggressive persecution of drug use over the well-being of students. Finally, after illegal entry, drug offenses are the most common cause of deportation, which breaks up families and ruins lives.

The most effective ways to address the overdose crisis are evidence-based public health and harm reduction approaches that take steps toward legalization. These include strategies like low-barrier supervised consumption services, drug checking, overdose prevention and response programs, safer supply initiatives, non-abstinence-based housing, naloxone distribution, and education on safer drug use (Harm Reduction International, 2020). Harm reduction strategies have rapidly expanded in Canada in efforts to curb overdose fatalities (Wallace & Pagan, 2019). Alongside them, efforts are currently being undertaken to implement supervised consumption sites in a number of U.S. cities (e.g., Seattle, New York, San Francisco) (Allyn, 2018), and a U.S. District Judge recently ruled that a bid to open a supervised injection site in Philadelphia does not violate the Controlled Substances Act, clearing the way for the operation of the first above-ground, supervised consumption venue in the United Sates (Levenson & del Valle, 2020).

These strategies save lives. This is partly because injection is more likely to have negative results when performed in an unsafe environment, using unregulated, impure substances, than when it occurs in safe context like a supervised injection site (Burris et al., 2004). Evidence from cohort and modeling studies in Canada suggests that safe injection sites are associated with lower overdose mortality, 67% fewer ambulance calls for treating overdoses, and a decrease in HIV infections (Ng et al., 2017). In fact, it is estimated that the recent scale-up of these treatment and harm reduction interventions in British Columbia prevented more than 3,000 potential overdose deaths between April 2016 and December 2017 (Irvine et al., 2019).

Yet despite the gradual uptake of harm reduction strategies, substance use treatment options, and public health approaches to drug use, such strategies have continued to struggle to reduce opioid overdoses and deaths, in part because of institutional, legal, and social barriers that prevent the large-scale application of such programs. What this indicates is a need not only for public health and harm reduction, but also for widespread institutional change. By scaling up safe supply initiatives and opioid distribution programs, we can take serious steps in the direction of effective drug regulation and legalization, putting an end to the harmful policies that have characterized our decades long "war" against drugs and the people who use them. Instead of waging war, we can opt for a health and human rights–based approach to drug use that makes opioids—and other drugs—safer and that values and dignifies the lives of the people who use them.

Broadening access to harm reduction services is especially important in light of the Covid-19 pandemic. The year 2020 marked the beginning of a parallel public health crisis that has since exposed a variety of systemic issues in public health and has illustrated just how deeply the drug war permeates health and social systems. Individuals who interact with these systems, including those who are in prison, homeless, immigrants, those with a substance use disorder, and those who rely on medication-assisted treatments, are unable to take the most basic of steps to avoid contracting and prevent the spread of Covid-19. This is because it is much more challenging for them to social-distance, to access medication-assisted treatment, and to receive other harm reduction resources, like clean needles or naloxone. Harm reduction program have been profoundly impacted by the Covid-19 pandemic, yet not enough attention has been paid to the vulnerability of people who live and work in these contexts. The need for the labor and expertise of PWUD in harm reduction programs has increased throughout the pandemic, especially since lockdowns and physical distancing measures have created new barriers to accessing harm reduction services and have interrupted the delivery of such services in many regions across the country. As Nancy Campbell (2019) and other have noted—and as has been the case in previous epidemics, like HIV—the specialized labor and expertise of PWUD are instrumental in adapting harm reduction programs and services to the new barriers that the Covid-19 pandemic has erected against preventing the harms associated with opioid overdose.

To that end, I want to conclude with a few additional implications that my analysis may have for the future of pain and people who use opioids.

First, in pain management and healthcare, it has become clear that what is most complicated, and perhaps most important about pain, is its subjective component. Thus, I hope that this initial attempt to trace a genealogy of pain management will spur additional research and a greater emphasis on the subjective aspect of pain. I hope it will push healthcare providers to consider—in addition to numerical scales—routinely incorporating multifaceted, qualitative evaluations of how a patient feels about their pain, and what they think it means (for their body, daily life, future, etc.), and the different kinds of needs and barriers that make it impossible to live "with" chronic pain.

Although it has been the case that regulatory discussions regarding opioids have, at times, explicitly addressed the question of pain and its management, they have not done so in ways that acknowledge pain's relationship to addiction, nor subjectivity, nor contextual specificity. Understanding the ways in which different understandings of pain have conditioned its treatment in medicine, culture, and politics could provide regulators with a clearer perspective regarding the lived experience of opioid use—a perspective that could help to mitigate some of the consequences that have emerged (and will continue to do so) under permissive regulatory logics. This book has argued that these logics, are defined by extreme uncertainty and strategic ignorance. And it has shown how they have played a huge role in conditioning the practices of opioid regulation and, correspondingly, the expansion of the gray market in which opioids often circulate.

One way of addressing extreme uncertainty in the context of opioid regulation would involve calling for new kinds of expertise to be incorporated into opioid-related policy discussions. As we have seen, the opioid epidemic has called into question several long-held assumptions about who and what constitute expertise in this domain. Increasingly, those actors who were previously seen as experts are being reframed as culprits: Pharmaceutical companies, regulators, physicians, and pain management specialists have all come under fire over the past several years for having fueled opioid addiction and overdoses among their patients and consumers. For these actors, the opioid epidemic seems to have amplified a growing crisis of legitimacy and widespread loss of trust. This is especially true among those who see the crisis as a disastrous consequence of physicians' bad decision-making, pharmaceutical greed, and the futility of our public health institutions. The cracks revealed in this current system of expertise call up a need for new kinds of experts and for their participation in drug policy discussions.

In line with drug scholars like Jarrett Zignon (2020) and Nancy Campbell (2019), both of whom make strong cases for the need to rethink expertise in drug policy, I argue that we need to provide a seat at the drug policy table for *lived* expertise, and that means providing seats to PWUD. After all, who would be better positioned to speak about the complex reality of pain and opioid use than people who use opioids themselves? Who better to help regulators understand the real trajectories of long-term opioid use than people in pain? Lived expertise, as it exists among individuals with different kinds of relationships to the drugs they take, has much to offer policymakers who are attempting to shrink the gray market and to reduce the harms inflicted upon PWUD by those actors and institutions who have—and continue to—fight a failed war on drugs.

As is written in Chapter 2, formal policy discussions about the risks and benefits of opioid-based pain management have lacked this kind of specialized and embodied knowledge. As such, regulators and policymakers have often been unable to predict exactly how the drugs they choose to send to market will be consumed by the people who need them. They could not anticipate, for example, just how many individuals who experience chronic pain would end up taking more OxyContin than "as prescribed" and how many of the abuse-deterrent mechanisms regulators approved to prevent problematic opioid use would end up sending it skyrocketing in new directions. They could not have known these things, in part, because they never asked or consulted the people who do know. PWUD, whose experiences are often rendered invisible in opioid regulation and drug policy, have been crucial actors in the application of harm reduction, acting as peers who both participate in and lead training in harm reduction practices like learning how to correctly identify an overdose or how to safely administer naloxone. But what we still lack is the meaningful inclusion of lived expertise in priority setting, research design and implementation, and policy discussions that lead to concrete measures and legislation geared toward saving human lives.

It is not enough to take a "compassionate approach" to drug use and PWUD if we do not center their experiences and voices in the process of achieving meaningful institutional change. We must fill the gaps in knowledge that characterize our drug policy and regulation landscape and transform extreme uncertainty into the production and circulation of knowledge gleaned from real experiences with pain, opioids, and PWUD.

Furthermore, the analysis of opioid regulation in Chapter 2 has had much to say about the strength of the ties between regulators and pharmaceutical companies. The role of these ties, as that chapter tried to show, has proven extremely detrimental to people who use opioids and has helped to engorge the pharmaceutical market with a surplus of prescription painkillers. Weakening the ties between the regulatory realm and big pharma by appointing commissioners with no personal or professional pharmaceutical connections or by removing companies and their proprietary systems from the landscape of postmarket surveillance could prove to be a productive means of dismantling the destructive alliance that has, tragically, helped lead us to where we are today.

On a more conceptual level, policymaking in this domain would benefit from rethinking its approach to pain and addiction in broader historical and cultural terms—as things that are not limited to their medical/scientific definitions and cannot be objectively measured and unilaterally treated, but as phenomena that have evolved unevenly within a matrix of cultural and institutional values, strategies, and interests. Put another way, if we are to address pain and addiction in a way that does not always hinge on opioids, we need to understand how the connection between pain and opioids developed in the first place. Thinking more extensively about the genealogies of pain management and addiction treatment may provide useful road maps that show us which routes we have already taken, and which might now be productively explored.

Pharmaceutical branding marks yet another domain in which this analysis may hold some implications for the future. First, it is worth considering restricting or eliminating altogether direct-to-consumer drug advertising (as is the case in nearly every other country in the world—countries that, it should also be noted, are not plagued by an opioid overdose crisis). Eliminating, restricting, or better regulating direct-to-consumer advertising, as Chapter 3's analysis of the branding of pain relief illustrates, may be an important step toward diminishing chronic patienthood and terminating the endless cycle of pharmaceutical optimization in which many of us find ourselves trapped. Specifically, regarding pain, regulating opioid-related advertising may be an important means of detaching pain from its warped relationship to the American dream.

With regard to opioid treatment, my analysis has noted some of the discrepancies in the quality of opioid addiction treatment among different demographic groups. The reality that certain "high quality of life" treatments, like buprenorphine, are less easily accessed by poor people and people of color is cause for considering the incorporation of buprenorphine as a Medicare drug. This should also be supplemented by offering free physician training and new guidelines that would loosen buprenorphine prescribing restrictions and broaden distribution networks. Doing so would help expand access to this drug for the people who need it most, with special attention paid to individuals living in poverty and to people of color, whose overdose deaths are now rising faster than ever.

These are but a few illustrations of some of the implications that could arise in the future, and to which the analysis developed in the previous pages points. But none of these proposed ameliorative actions alone are going to solve the growing crisis we have on our hands. And despite being hopeful, I sometimes cannot help but wonder: Is it possible to wake up from this American nightmare? Or are we destined to hit rock bottom (whatever that means) before that finally happens? Though these are haunting questions, I do believe that with closer attention paid to the analysis and implications suggested in the chapters, along with the crucial recognition that everyone's pain matters and warrants the best treatment society can offer, we stand a fighting chance.

Finally, I want to end this book with one last suggestion—that we all begin to rethink pain and addiction not as individual problems but as problems that can only be understood in complex, social, and cultural terms. This is a task that would not only benefit PWUD, their family members, policymakers, and others but could also serve as a crucial agenda for future research. Expanding on the social and cultural study of pain means investing in developing a better understanding of this complex phenomenon beyond the sanitized contexts of the hospital, clinic, and laboratory, and trying to come to terms with the ways in which pain shapes and is shaped by historical, social, and political dynamics. It requires that we begin to think more broadly and more critically about pain not solely in terms of what it is and how to measure it, but in terms of how it functions as a part of other key cultural processes. Examining the *how* of pain (how it is defined, realized, and assessed; how we relate to our own pain and the pain of others; and how pain becomes represented and intervened upon in a variety of domains) might just help us make sense of the turbulence that—being both affective and material, individual and collective—colors the landscape of our present moment. The United States is, in many ways, a nation in pain. To heal our wounds, we must begin by thinking harder about how it is that we have come to hurt.

References

ABC News. (2017, October 30). The new face of heroin addiction. https://www.youtube.com/watch?v=cskq_zGVSZs

Abraham, J., & Davis, C. (2009). Drug evaluation and the permissive principle: Continuities and contradictions between standards and practices in antidepressant regulation. *Social Studies of Science, 39*(4), 569–598.

Acker, C.J. (2002). *Creating the American junkie: Addiction research in the classic era of narcotic control.* Baltimore: Johns Hopkins University Press.

Adams, J., Bledsoe, G.H., & Armstrong, J.H. (2016). Are pain management questions in patient satisfaction surveys driving the opioid epidemic? *American Journal of Public Health, 106*(6), 985–986. https://doi.org/10.2105/AJPH.2016.303228

Administrative Office of the U.S. Courts. (2004, June). Sentencing commission takes new look at mandatory minimums. *Third Brand News.* https://web.archive.org/web/20121011112132/http://www.uscourts.gov/News/TheThirdBranch/10-06-01/Sentencing_Commission_Takes_New_Look_at_Mandatory_Minimums.aspx

Administrative Office of the U.S. Courts. (2007, June 26). Mandatory minimum terms result in harsh sentencing. Archived from the original on December 8, 2010. https://web.archive.org/web/20101208070157/http://www.uscourts.gov/News/NewsView/07-06-26/Mandatory_Minimum_Terms_Result_In_Harsh_Sentencing.aspx

Administrative Office of the U.S. Courts. (2010, June 1). Sentencing commission takes new look at mandatory minimums. *Third Branch News.* https://web.archive.org/web/20121011112132/http://www.uscourts.gov/News/TheThirdBranch/10-06-01/Sentencing_Commission_Takes_New_Look_at_Mandatory_Minimums.aspx

Ahmad, F.B., Cisewski, J.A., Rossen, L.M., & Sutton, P. (2022). Provisional drug overdose death counts. National Center for Health Statistics. Retrieved from https://www.cdc.gov/nchs/nvss/vsrr/drug-overdose-data.htm

Alexander, M. (2010/2012). *The new Jim Crow: Mass incarceration in the age of colorblindness* (Rev. ed.). New York: New Press.

Allyn, B. (July 2018). Cities planning supervised drug injection sites fear justice department reaction. *NPR.* https://www.npr.org/sections/health-shots/2018/07/12/628136694/harm-reduction-movement-hits-obstacles

American Medical Association House of Delegates. (2000). Resolution 514. https://www.ama-assn.org/system/files/a22-514.pdf

American Medical Directors Association. (1999). *Chronic pain management in the long-term care setting.* Columbia, MD: American Medical Directors Association.

Anderson, K.O., Green, C.R., & Payne, R. (2009). Racial and ethnic disparities in pain: Causes and consequences of unequal care. *Journal of Pain, 10*(12), 1187–1204. https://doi.org/10.1016/j.jpain.2009.10.002

Andrews, T.M. (2017, February 4). Louisville officials receive 52 overdose calls in 32 hours as country's addiction epidemic explodes. *Washington Post.* https://www.washingtonpost.com/news/morning-mix/wp/2017/02/14/louisville-officials-receive-52-overd ose-calls-in-32-hours-as-countrys-addiction-epidemic-explodes/?utm_term=.74e2b b2824b7

Angell, M. (2004). *The truth about drug companies: How they deceive us and what to do about it.* New York: Random House.

Armstrong, D. (1995). The rise of surveillance medicine. *Sociology of Health & Illness, 17*(3), 393–404.

Armstrong, D. (2016, September 22). Secret trove reveals bold "crusade" to make OxyContin a blockbuster. *STAT.* https://www.statnews.com/2016/09/22/abbott-oxycontin-crusade/

ASAM. (2001). Definitions related to the use of opioids for the treatment of pain: Consensus statement of the American Academy of Pain Medicine, the American Pain Society, and the American Society of Addiction Medicine. Retrieved from https://pubmed.ncbi.nlm.nih.gov/11579797/

ASAM. (2013). Advancing access to addiction medications: Implications for opioid addiction treatment. Retrieved from https://www.asam.org

ASAM. (2016). Opioid addiction: 2016 facts and figures. http://www.asam.org/docs/default-source/advocacy/opioid-addiction-disease-facts-figures.pdf

Associated Press. (2007, February 28). Ex-FDA chief gets probation, fine for lying about stocks. *The Washington Post.* http://www.washingtonpost.com/wp-dyn/content/article/2007/02/27/AR2007022701521.html

Ballantyne, J.C., & Sullivan, M.D. (2015). Intensity of chronic pain—The wrong metric? *New England Journal of Medicine, 373,* 2098–2099. https://doi.org/10.1056/NEJMp1507136

Bandura, A. (1977). Self-efficacy: Toward a unifying theory of behavioral change. *Psychology Review, 84,* 191–215.

Banet-Weiser, S. (2012). *Authentic^{TM}: The politics of ambivalence in a brand culture.* New York: New York University Press.

Baszanger, I. (1998). *Inventing pain medicine: From the laboratory to the clinic.* New Brunswick, NJ: Rutgers University Press.

Bebinger, M. (2019, March 21). Fentanyl-linked deaths: The U.S. opioid epidemic's third wave begins. *NPR.* https://www.npr.org/sections/health-shots/2019/03/21/704557684/fentanyl-linked-deaths-the-u-s-opioid-epidemics-third-wave-begins

Beck, U. (1992). *Risk society: Towards a new modernity.* London: Sage Publications.

Beletsky, L., & Davis, C. (2017). Today's fentanyl crisis: Prohibition's iron law, revisited. *International Journal of Drug Policy, 46,* 156–159.

Beletsky, L., Seymour, S., Kang, S., Sigel, Z., Sinha, M.S., Marino, R., Dave, A., & Freifeld, C. (2020). Fentanyl panic goes viral: The spread of misinformation about overdose risk from casual contact with fentanyl in mainstream and social media. *International Journal of Drug Policy, 86,* 102951.

Beltran, R. (2005). The gold standard: The challenge of evidence-based medicine and standardization in health care. *Journal of the National Medical Association, 97*(1), 110.

Benner, K. (2020). Purdue pharma pleads guilty to role in opioid crisis as part of deal with Justice Dept. *The New York Times.* https://www.nytimes.com/2020/11/24/us/politics/purdue-pharma-opioids-guilty-settlement.html

Bernstein, L. (2015, December 18). Deaths from opioid overdoses set a record in 2014. *Washington Post.* https://www.washingtonpost.com/news/to-your-health/wp/2015/12/11/deaths-from-heroin-overdoses-surged-in-2014/?utm_term=.1803a1c4b9ef

REFERENCES 163

Bever, L. (2018, September 8). The man who made billions of dollars from OxyContin is pushing a drug to wean addicts off opioids. *The Washington Post.* https://www.washingtonpost.com/news/business/wp/2018/09/08/the-man-who-made-billions-of-dollars-from-oxycontin-is-pushing-a-drug-to-wean-addicts-off-opioids/?noredirect=on&utm_term=.4ce9452f3e2b

Bode, A.M., & Dong, Z. (2010). Cancer prevention research—then and now. *National Review of Cancer, 9*(7), 508–516. https://doi.org/10.1038/nrc2646

Bonica, J. (1953/1990). *The management of pain* (2nd ed.). Lea & Febiger

Bourdet, K. (2012, September 18). How big pharma hooked America on legal heroin. *Vice.* http://motherboard.vice.com/read/how-big-pharma-hooked-america-on-legal-heroin

Bourgois, P.I., & Schonberg, J. (2009). *Righteous dopefiend.* Berkeley: University of California Press.

Brandt, A. (2009). *The cigarette century: The rise, fall, and deadly persistence of the product that defined America.* New York: Basic Books.

Brena, S., Chapman, S.L., & Decker, R. (1981). Chronic pain as a learned experience: Emory University Pain Control Center. *NIDA Research Monograph, 36,* 76–83.

Burris, S., Blankenship, K.M., & Donoghoe, M. (2004). Addressing the "risk environment" for injection drug users: The mysterious case of the missing cop. *Milbank Quarterly, 82*(10), 125–156.

Burroughs, W.S. (1953/2003). *Junky: The definitive text of "junk."* New York, NY: Penguin Group.

Burroughs, W.S. (1959/2013). *Naked lunch: The restored text.* New York, NY: Grove Press.

Bush, J. (2016, January 4). Addressing the heartbreak of addiction. *Medium.* https://medium.com/@JebBush/addressing-the-heartbreak-of-addiction-53854d26f8d3

Calabressi, M. (2015, June 15). The price of relief: Why America can't kick its painkiller problem. *TIME.* https://time.com/3908648/why-america-cant-kick-its-painkiller-problem/

Calcaterra, S., Glanz, J., & Binswanger, I.A. (2013). National trends in pharmaceutical opioid related overdose deaths compared to other substance related overdose deaths: 1999–2009. *Drug and Alcohol Dependence, 131*(3), 263–270. https://doi.org/10.1016/j.drugalcdep.2012.11.018

Campbell, J. (1996, November 11). *Pain as the 5th vital sign [presidential address].* American Pain Society News. https://fbaum.unc.edu/teaching/articles/Campbell1996Pain.pdf

Campbell, N.D. (2000). *Using women: Gender, drug policy, and social justice.* New York: Routledge.

Campbell, N.D. (2007). *Discovering addiction: The science and politics of substance abuse research.* Ann Arbor: University of Michigan Press.

Campbell, N. (2010). Toward a critical neuroscience of "addiction." *Biosocieties, 5*(1), 89–104.

Campbell, N.D., & Lovell, A.M. (2012). The history of the development of buprenorphine as an addiction therapeutic: Campbell & Lovell. *Annals of the New York Academy of Sciences, 1248*(1), 124–139. https://doi.org/10.1111/j.1749-6632.2011.06352.x

Carpenter, D. (2010). *Reputation and power: Organizational image and pharmaceutical regulation at the FDA* (Princeton Studies in American Politics: Historical, International, and Comparative Perspectives) [Kindle ed.]. Princeton, NJ: Princeton University Press.

Carr, E.S. (2013). Signs of sobriety: Rescripting American addiction counseling. In E. Raikhel & W. Garriott (Eds.), *Addiction trajectories* (pp. 160–167). Durham, NC: Duke University Press.

Carrico, J.A., Mahoney, K., Raymond, K.M., Mims, L., Smith, P.C., Sakai, J.T., Mikulich-Gilbertson, S.K., Hopfer, C.J., & Bartels, K. (2018). The association of patient satisfaction-based incentives with primary care physician opioid prescribing. *Journal of the American Board of Family Medicine, 6*, 941–943.

Carroll, L. (2014, April 8). Hooked: A teacher's addiction and the new face of heroin. *Today*. https://www.today.com/health/hooked-teacher-s-addiction-new-face-heroin-t74881

Case, A., & Deaton, A. (2015). Rising morbidity and mortality in midlife among white non-Hispanic Americans in the 21st century. *Proceedings of the National Academy of Sciences of the United States of America, 112*(49), 15078–15083. http://www.pnas.org/content/112/49/15078

Case, A., & Deaton, A. (2017, March 23). Mortality and morbidity in the 21st century. Brookings Papers on Economic Activity. https://www.brookings.edu/bpea-articles/mortality-and-morbidity-in-the-21st-century/

Caulkins, J.P. (1997, January 1). *Are mandatory minimum drug sentences cost-effective?* RAND Corporation. https://www.rand.org/pubs/research_briefs/RB6003.html

CDC. (2011a). Vital signs: Overdoses of prescription opioid pain relievers—United States, 1999–2008. *Morbidity and Mortality Weekly Report, 60*(43), 1487–1492. www.cdc.gov/mmwr/preview/mmwrhtml/mm6043a4.htm

CDC. (2011b). Press release: Prescription painkiller overdoses at epidemic levels. https://www.cdc.gov/media/releases/2011/p1101_flu_pain_killer_overdose.html

CDC. (2015). *Wide-ranging online data for epidemiologic research (WONDER)*. Atlanta, GA: CDC, National Center for Health Statistics. Retrieved from http://wonder.cdc.gov

CDC. (2016). Synthetic opioid overdose data. Retrieved from https://www.cdc.gov/drugoverdose/deaths/synthetic/2016-2017.html

CDC. (2017a). Drug overdose deaths in the United States, 1999–2016. CDC NCHS Data Brief No. 294. Retrieved from https://www.cdc.gov/nchs/products/databriefs/db294.htm

CDC. (2017b). Multiple cause of death, 1999–2016 [data set]. Retrieved from https://wonder.cdc.gov/mcd-icd10.html

CDC. (2017c). Provisional counts of drug overdose deaths, as of 8/6/2017. National Center for Health Statistics. National Vital Statistics System. Retrieved from https://www.cdc.gov/nchs/data/health_policy/monthly-drug-overdose-death-estimates.pdf

CDC. (2017d). Understanding the epidemic. Retrieved from https://www.cdc.gov/drugoverdose/epidemic/index.html

CDC. (2019). Still not enough naloxone where it's most needed [Press release]. *CDC*. https://www.cdc.gov/media/releases/2019/p0806-naloxone.html

CDC. (2021). Understanding the epidemic. Retrieved from https://www.cdc.gov/drugoverdose/epidemic/index.html

CDC/NCHS. (2018, January 11). *National vital statistics system, mortality*. CDC WONDER database. Atlanta, GA: US Department of Health and Human Services, CDC. Retrieved from https://wonder.cdc.gov/

Center for Public Integrity (CPI). (2016, September 18). Pharma lobbying held deep influence over opioid policies. https://publicintegrity.org/state-politics/pharma-lobbying-held-deep-influence-over-opioid-policies/

Center for Responsive Politics. (2017). Client profile: Pharmaceutical Research and Manufacturers of America. Retrieved from https://www.opensecrets.org/lobby/client sum.php?id=D000000504&year=2017.

Chabal, C., Jacobson, L., Mariano, A.J., & Chaney, E. (1998). Opiate abuse or undertreatment? *Clinical Journal of Pain, 14*(1), 90–91.

Chatterjee, P., Joynt, K.E., Orav, E.J., Jha, & A.K. (2012). Patient experience in safety-net hospitals: Implications for improving care and value-based purchasing. *Archives of Internal Medicine, 172*(16), 1204–1210. https://doi.org/10.1001/archinternmed.2012.3158

Checklist: 11 ways to ensure proper pain management. (2004, June 10). Retrieved from https://web.archive.org/web/20040610235744/http://partnersagainstpain.com/index-pc.aspx?sid=29&aid=7686

Cherlin, A. (2018). Psychological health and socioeconomic status among non-Hispanic whites. *PNAS, 115*(28), 7176–7178.

Christensen, J. (2017, June 8). FDA wants opioid painkiller pulled off market. *CNN.* http://www.cnn.com/2017/06/08/health/fda-opioid-opana-er-bn/index.html

Cicero, T.J., & Ellis, M.S. (2015). Abuse-deterrent formulations and the prescription opioid abuse epidemic in the United States: Lessons learned from OxyContin. *JAMA Psychiatry, 72*(5), 424–429.

Cicero, T.J., Ellis, M.S., Surratt, H.L., Kurtz, & S.P. (2014). The changing face of heroin use in the United States: A retrospective analysis of the past 50 years. *JAMA Psychiatry, 71*(7), 821–826.

Cicero, T.J., Ellis, M.S., & Surratt, H.L. (2012). Effect of abuse-deterrent formulation of OxyContin. *New England Journal of Medicine, 367*(2), 187–189. https://doi.org/10.1056/NEJMc1204141

Cicero, T., Inciardi, J., & Munoz, A. (2005). Trends in abuse of OxyContin and other opioid analgesics in the United States: 2002–2004. *Journal of Pain, 6,* 662–672.

Cohen, S. (2016, August 13). Prosecution trend: After fatal OD, dealer charged with death. *Associated Press.* https://apnews.com/f1f28eca1b274e8f88f218bbd7d181d6/prosecution-trend-after-fatal-od-dealer-charged-death

Collins, M., & Vakharia, S.P. (2020). Criminal justice reform in the fentanyl era: One step forward, two steps back. New York: Drug Policy Alliance. https://drugpolicy.org/sites/default/files/dpa-cj-reform-fentanyl-era-v.3_0.pdf

Committee on Education and the Workforce. (2018, February 15). Joint subcommittee hearing examines the impact of opioids in the workplace [press release]. https://edworkforce.house.gov/news/documentsingle.aspx?DocumentID=402497

Compton, W.M., Jones, C.M., & Baldwin, G.T. (2016). Relationship between nonmedical prescription-opioid use and heroin use. *New England Journal of Medicine, 374*(2), 154–163.

Congressional Record. (1999a). Hearing before the subcommittee on health and environment of the Committee in Commerce, House of Representatives. Congressional Record - House (106th Congress 1st Session): 1–23.

Congressional Record. (1999b). Drug Addiction Treatment Act of 1999. Congressional Record - Senate (106th Congress): S1089–S1093.

Congressional Record. (2000). Drug Addiction Treatment Act of 2000. Congressional Record - Senate (106th Congress): S9111.

Conrad, P. (2007). *The medicalization of society: On the transformation of human conditions into treatable disorders.* Baltimore: Johns Hopkins University Press.

Conrad, P., & Muñoz, V.L. (2010). The medicalization of chronic pain. In M. Moeller & L. Gormsen, *The role of chronic pain and suffering in contemporary society* (pp. 13–24). Aarhus, Denmark: Tidsskrift for Forskning i Sygdom og Samfund.

Conrad, P., & Schneider, J.W. (1980/1992). *Deviance and medicalization: From badness to sickness: With a new afterword by the authors* (Expanded ed.). Philadelphia: Temple University Press.

Controlled Substances Act of 1970 (CSA). (1970). Pub. L. No. 91-513, 84 Stat. 1242; *Drug Enforcement Administration, physician's manual: An informational outline of the Controlled Substances Act of 1970*. (1978). Washington, DC: U.S. Department of Justice.

Courtwright, D.T. (2001). *Dark paradise: A history of opiate addiction in America* (Enlarged ed.). Cambridge, MA: Harvard University Press.

Craig, K.D., Holmes, C., Hudspith, M., Moor, G., Moosa-Mitha, M., Varcoe, C., & Wallace, B. (2020). Pain in persons who are marginalized by social conditions. *Pain*, *161*(2), 261–265. https://doi.org/10.1097/j.pain.0000000000001719

Crews, J.C., & Denson, D.D. (1990). Recovery of morphine from a controlled-release preparation: A source of opioid abuse. *Cancer*, *66*, 2642–2644.

Daemmrich, A.A. (2004). *Pharmacopolitics: Drug regulation in the United States and Germany*. Chapel Hill: The University of North Carolina Press.

Dahl, J. (1993). State cancer pain initiatives. *Journal of Pain and Symptom Management*, *8*(6), 372–375.

Dahl, J. (1997, April 20). Painkiller myths. *The New York Times*.

Dahlhamer, J., Lucas, J., Zelaya, C., Nahin, R., Mackey, S., DeBar, L., Kerns, R., Von Korff, M., Porter, L., & Helmick, C. (2018). Prevalence of chronic pain and high-impact chronic pain among adults—United States, 2016. *Morbidity and Mortality Weekly Report (MWWR)*, *67*(36), 1001–1006. https://www.cdc.gov/mmwr/volumes/67/wr/mm6736a2.htm

Daubresse, M., Chang, H., Yu, Y., Viswanathan, S., Shah, N.D., Stafford, R.S., Kruszewski, S.P., & Alexander, G.C. (2013). Ambulatory diagnosis and treatment of nonmalignant pain in the United States, 2000–2010. *Medical Care*, *51*(10), 870–878.

Davis, C.S., Green, T., & Beletsky, L. (2017). Action, not rhetoric, needed to reverse the opioid overdose epidemic. *The Journal of Law, Medicine & Ethics*, *45*(S1), 20–23.

DEA. (2001). Drugs and chemicals of concern: Summary of medical examiner reports on Oxycodone-related deaths. Office of National Drug Control Policy. Pulse Check Report. Retrieved from http://www.health.org/govpubs/dea/oxydeaths.htm

DEA. (2016a). Carfentanil: A dangerous new factor in the U.S. opioid crisis [Officer Safety Alert]. Retrieved from https://www.dea.gov/press-releases/2016/09/22/dea-issues-carfentanil-warning-police-and-public

DEA. (2016b, September 22). DEA issues carfentanil warning to police and public. National Media Affairs Office. Retrieved from https://www.dea.gov/press-releases/2016/09/22/dea-issues-carfentanil-warning-police-and-public

DEA. (2017, June). Fentanyl: A briefing guide for first responders. Retrieved from http://dig.abclocal.go.com/wls/documents/DEA_Fentanyl_Publication.pdf

DeJong, C., Aguilar, T., Tseng, C-W., Lin, G.A., Boscardin, W.J., & Dudley, R.A. (2006). Pharmaceutical industry-sponsored meals and physician prescribing patterns for Medicare beneficiaries. *JAMA Internal Medicine*, *176*(8), 1114–1110. https://doi.org/10.1001/jamainternmed.2016.2765

Deprez, E.E., & Barrett, P.M. (2017, October 5). The lawyer who beat Big Tobacco takes on the opioid industry. *Bloomberg Businessweek* https://www.bloomberg.com/news/featu res/2017-10-05/the-lawyer-who-beat-big-tobacco-takes-on-the-opioid-industry

Derkatch, C. (2018). The self-generating language of wellness and natural health. *Rhetoric of Health & Medicine, 1*(1–2), 132–160.

DOJ. (2001). *Information bulletin: OxyContin diversion and abuse.* Johnstown, PA: National Drug Intelligence Center.

Dole, V.P., & Nyswander, M. (1967). Heroin addiction - A metabolic disease. *Archives of Internal Medicine, 120*(1), 19–24.

Doleac, Jennifer L., & Mukherjee, Anita. (2021). The effects of naloxone access laws on opioid abuse, mortality, and crime. *SSRN.* https://ssrn.com/abstract=3135264 or http://dx.doi.org/10.2139/ssrn.3135264

Donovan, E.E., Brown, L.E., & Crook, B. (2015). Patient satisfaction with medical disclosure and consent documents for treatment: Applying conceptualizations of uncertainty to examine successful attempts at communicating risk. *Journal of Communication in Healthcare: Strategies, Media and Engagement in Global Health, 8*(3), 220–232. https://doi.org/10.1179/1753807615Y.0000000008

Dowell, D., Kunins, H.V., Farley, T.A. (2013). Opioid analgesics—Risky drugs, not risky patients. *JAMA, 309*(21), 2219–20.

Drug Policy Alliance. (2019). *Rethinking the "drug dealer."* New York: Drug Policy Alliance. https://drugpolicy.org/sites/default/files/dpa-rethinking-the-drug-dealer_0.pdf

Drug Policy Alliance. (2021). Uprooting the drug war. Retrieved from https://uprootin gthedrugwar.org

Dumit, J. (2004). *Picturing personhood: Brain scans and biomedical identity.* Princeton, NJ: Princeton University Press.

Dumit, J. (2012). *Drugs for life: How pharmaceutical companies define our health.* Durham, NC: Duke University Press.

Duster, T. (1970). *The legislation of morality.* New York: The Free Press.

DVA. (2000). Pain as the 5th vital sign toolkit. Washington, DC: Geriatrics and Extended Care Strategic Healthcare Group, National Pain Management Coordinating Committee, Veterans Health Administration.

DVA. (2017). America's wars. Retrieved from https://www.census.gov/history/pdf/va-veterans-62017.pdf

Easterbrook, G. (2014, May 26). Painkillers, NFL's other big problem. *ESPN.* http://www.espn.com/nfl/story/_/id/10975522/excerpt-painkillers-abuse-nfl-king-sports-gregg-easterbrook

Eddy, D. (1982). 18 - probabilistic reasoning in clinical medicine: Problems and opportunities. In A. Tversky, *Judgement under uncertainty: Heuristics and biases.* (pp. 249–267). Cambridge: Cambridge University Press.

Eddy, D. (1984). Variations in physician practice: The role of uncertainty. *Health Affairs, 3*(2), 74–89.

Eddy, D. (1988). The quality of medical evidence: Implications for quality of care. *Health Affairs, 7*(1), 19–32.

Eddy, D. (1990). Practice policies: Guidelines for methods. *Journal of the American Medical Association, 263*(3), 1839–1841.

Eddy, N.B. (Ed.). (1973). *The National Research Council involvement in the opiate problem, 1928-1971.* Washington, DC: National Academy of Sciences.

Egoscue, Inc. The Egoscue method. Retrieved from www.egoscue.com

168 REFERENCES

Eligon, J., & Kovaleski, S.F. (2016, June 2). Prince died from accidental overdose of opioid painkiller. *The New York Times*. https://www.nytimes.com/2016/06/03/arts/music/prince-death-overdose-fentanyl.html?_r=0

Elliot, C. (2003). *Better than well: American medicine meets the American dream*. New York: W.W. Norton and Company, Inc.

Ensuring that death and serious injury are more than a business cost: OxyContin and defective products. (2007). Committee on the Judiciary, Senate, 3 (testimony of John L. Brownlee). https://www.judiciary.senate.gov/imo/media/doc/Brownlee%20Testimony%20073107.pdf

Epstein, S. (1996). *Impure science: AIDS, activism, and the politics of knowledge*. Berkeley: University of California Press.

Falkenberg, K. (2013, January 2). Why rating your doctor is bad for your health. *Forbes*, http://www.forbes.com/sites/kaifalkenberg/2013/01/02/why-rating-your-doctor-is-bad-for-your-health/#4015ff072f15

Families Against Mandatory Minimums (FAMM). (2023). Our work. http://famm.org/projects/federal/

FDA. (2005a). Guidance for industry: The development and use of RiskMAPs. https://www.fda.gov/media/71268/download

FDA. (2005b). Guidance for industry: Good pharmacovigilance practices and pharmacoepidemiologic assessment. https://www.fda.gov/files/drugs/published/Good-Pharmacovigilance-Practices-and-Pharmacoepidemiologic-Assessment-March-2005.pdf

FDA. (2010, July 23). *Joint meeting of the Anesthetic and Life Support Drug Advisory Committee (ALSDAC) & Drug Safety and Risk Management Committee (DSaRM)*. Adelphi, MD: Center for Drug Evaluation and Research.

FDA. (2013, February 6). FDA's efforts to address the misuse and abuse of opioids. Retrieved from https://www.fda.gov/Drugs/DrugSafety/InformationbyDrugClass/ucm337852.htm

FDA. (2015). Abuse-deterrent opioids—Evaluation and labeling. Guidance for industry. Silver Spring, MD: U.S. Department of Health and Human Services. Food and Drug Administration. Center for Drug Evaluation and Research (CDER). https://www.fda.gov/media/84819/download

FDA. (2016). FDA Opioids Action Plan [Fact sheet]. Retrieved from https://www.fda.gov/drugs/information-drug-class/fda-opioids-action-plan

FDA. (2017a, June 8). FDA requests removal of Opana ER for risks related to abuse [FDA news release]. https://www.fda.gov/NewsEvents/Newsroom/PressAnnouncements/ucm562401.htm?source=govdelivery&utm_medium=email&utm_source=govdelivery andhttps://www.fda.gov/news-events/press-announcements/fda-requests-removal-opana-er-risks-related-abuse

FDA. (2017b, November 30). Statement from FDA Commissioner Scott Gottlieb, M.D., on the approval of a new formulation of buprenorphine and FDA's efforts to promote more widespread innovation and access to opioid addiction treatments. *PR Newswire*. https://www.prnewswire.com/news-releases/statement-from-fda-commissioner-scott-gottlieb-md-on-the-approval-of-a-new-formulation-of-buprenorphine-and-fdas-efforts-to-promote-more-widespread-innovation-and-access-to-opioid-addiction-treatments-300564874.html

FDA. (2021). Center for drug evaluation and research: Drug safety priorities - 2021. Silver Spring, MD: FDA. https://www.fda.gov/media/155795/download

REFERENCES 169

FDA. (n.d.). Timeline of selected FDA activities & significant events addressing opioid misuse and abuse. Retrieved from https://www.fda.gov/Drugs/DrugSafety/InformationbyDrugClass/ucm338566.htm

FDA. (2017). FDA facts: Abuse-deterrent opioid medications. Retrieved from https://www.fda.gov/newsevents/newsroom/factsheets/ucm514939.htm

FDA. (2015, December 12). FDA guidance for industry: Opioids with abuse-deterrent properties. Retrieved from https://web.archive.org/web/20151119030738/http://www.teamagainstopioidabuse.com:80/fda-guidance-opioids-with-abuse-deterrent-properties/

Federation of State Medical Boards of the United States (FSMB). (1988). *A guide to the essentials of a modem medical practice act* (5th ed.). Forth Worth, TX: FSMB.

Fernandes, C.O., Melo, C., Besset, V.L., & Bicalho, P.P. (2016). Biopolitics and pain: Approximations between Foucault and Lacanian psychoanalysis. *Psico-USF, 21*(1), 189–196.

Find a doctor. (2004, June 10). *Partners Against Pain.* Retrieved from https://web.archive.org/web/20040610133321/http://partnersagainstpain.com/index-pc.aspx?sid=15

Fink, D.S., Schleimer, J.P., Sarvet, A., Grover, K.K., Delcher, C., Castillo-Carniglia, A., Kim, J.H., Rivera-Aguirre, A.E., Henry, S.G., Martins, S.S., & Cerdá, M. (2018). Association between prescription drug monitoring programs and nonfatal and fatal drug overdoses: A systematic review. *Annals of Internal Medicine.* Advance online publication. https://pubmed.ncbi.nlm.nih.gov/29801093/

Fishbain, D.A., Rosomoff, H.L., & Rosomoff, R.S. (1992). Drug abuse, dependence, and addiction in chronic pain patients. *Clinical Journal of Pain, 8*(2), 77–85. https://doi.org/10.1097/00002508-199206000-00003. PMID: 1633386.

Foley, K. (1985). The treatment of cancer pain. *New England Journal of Medicine, 313,* 84–95.

Fordyce, W., Roberts, A.H., & Sternbach, R.A. (1985). The behavioral management of chronic pain: A response to critics. *Pain, 22,* 113–125.

Foster, G., Taylor, S.J.C., Eldridge, S., Ramsay, J., & Griffiths, C.J. (2007). Self-management education programmes by lay leaders for people with chronic conditions. *Cochrane Database of Systematic Reviews,* Article CD005108.

Foucault, M. (1977a). *Discipline and punish: The birth of the prison.* New York, NY: Random House LLC.

Foucault, M. (1977b). The confession of the flesh [Interview]. In C. Gordon (Ed.). (1980), *Power/knowledge: Selected interviews and other writings* (pp. 194–122). New York: Pantheon Books.

Foucault, M. (1991). Governmentality. In G. Burchell, C. Gordon, & P. Miller (Eds.), *The Foucault Effect: Studies in governmentality* (pp. 87–104). Chicago: University of Chicago Press.

Foucault, M. (2008). *The birth of biopolitics: Lectures at the Collège de France, 1978–1979.* New York: Picador.

Frasco, P.E., Sprung, J., & Trentmon, T.L. (2005). The impact of the JCAHO pain initiative on perioperative opiate consumption and recovery room length of stay. *Anesthesia and Analgesia, 100,* 162–168.

Friedman, J., Kim, D., Schneberk, T., Bourgois, P., Shin, M., Celious, A., & Schriger, D.L. (2019). Assessment of racial/ethnic and income disparities in the prescription of opioids and other controlled medications in California. *JAMA Internal Medicine, 179*(4), 469–476. https://doi.org/10.1001/jamainternmed.2018.6721

Frostenson, S. (2017 April 7). The crackdown on opioid prescriptions is leaving chronic pain patients in limbo. *Vox.* https://www.vox.com/science-and-health/2017/4/7/14738292/crackdown-opioid-prescriptions-chronic-pain

Frydl, K. (2017, February 5). How drug reformers have failed the opioid crisis. *Medium.* https://medium.com/@kfrydl/how-drug-reformers-have-failed-the-opioid-crisis-97000b869899

Fuqua, J.V. (2012). *Prescription TV: Therapeutic discourse in the hospital and at home.* Durham, NC: Duke University Press.

Ganim, S. (2017, February 16). China's fentanyl ban a "game changer" for opioid epidemic, DEA officials say. *CNN.* http://www.cnn.com/2017/02/16/health/fentanyl-china-ban-opioids/

GAO. (2003). *Prescription drugs: OxyContin abuse and diversion and efforts to address the problem.* Report number GAO-04-110. Washington, DC: U.S. General Accounting Office.

Gilson, A.M., & Joranson, D.E. (2001). Controlled substances and pain management: Changes in knowledge and attitudes of state medical regulators. *Journal of Pain and Symptom Management, 21*(3), 227–237.

Gladwell, M. (1992, April 10). FDA implements changes in drug approval process. https://www.highbeam.com/doc/1P2-999993.html

Goodnough, A. (2016, May 11). Treating pain without feeding addiction at epicenter for opioids. *The New York Times.* https://www.nytimes.com/2016/05/12/us/opioids-addiction-chronic-pain.html

Gottlieb, S. (2007). Drug safety proposals and the intrusion of federal regulation into patient freedom and medical practice. *Health Affairs, 26*(3), 664–677. http://content.healthaffairs.org/content/26/3/664.full

Graham, C. (2017). *Happiness for all? Unequal hopes and lives in pursuit of the American Dream.* Princeton, NJ: Princeton University Press.

Green, C.R., Baker, T.A., Sato, Y., Washington, T.L., & Smith, E.M. (2003). Race and chronic pain: A comparative study of young black and white Americans presenting for management. *Journal of Pain, 4*(4), 176–183.

Greene, M.S., & Chambers, R.A. (2015). Pseudoaddiction: Fact or fiction? An investigation of the medical literature. *Current Addiction Reports, 2*(4), 310–317.

Griffin III, O.H., & Spillane, J.F. (2012). Pharmaceutical regulation failures and changes: Lessons learned from OxyContin abuse and diversion. *Journal of Drug Issues, 43*(2), 164–175.

Gudin, J.A., Mogali, S., Jones, J.D., & Comer, S.D. (2013). Risks, management, and monitoring of combination opioid, benzodiazepines, and/or alcohol use. *Postgraduate Medicine, 125*(4), 115–130. https://doi.org/10.3810/pgm.2013.07.2684

Guitar-opana-man. (2008, November 29). *The best way to take Opana ER* [Online forum comment]. Bluelight. http://www.bluelight.org/vb/threads/406641-Opana-ER

Gupta, R., Shah, N.D., & Ross, J.S. (2016). The rising price of naloxone – risks to efforts to stem overdose deaths. *New England Journal of Medicine, 375,* 2213–2215.

Gusovsky, Dina. (2015, November 4). The pain killer: A drug company putting profits above patients. *CNBC.* https://www.cnbc.com/2015/11/04/the-deadly-drug-appeal-of-insys-pharmaceuticals.html

Hall, S. (1986). Gramsci's relevance for the study of race and ethnicity. *Journal of Communication Inquiry, 10*(2), 5–27. https://doi.org/10.1177/019685998601000202

REFERENCES 171

Halpern, L.M., & Robinson, J. (1985). Prescribing practices for pain in drug dependence: A lesson in ignorance. *Advances in Alcohol & Substance Abuse*, 5(1–2), 135–162. https://doi.org/10.1300/J251v05n01_10

The Hamilton Project. (2016). Rates of drug use and sales, by race; rates of drug related criminal justice measures, by race. Retrieved from https://www.hamiltonproject.org/charts/rates_of_drug_use_and_sales_by_race_rates_of_drug_related_criminal_justice

Hansen, H., & Roberts, S.K. (2012). Two tiers of biomedicalization: Methadone, buprenorphine, and the racial politics of addiction treatment. In J. Netherland (Ed.), *Advances in Medical Sociology* (Vol. 14, pp. 79–102). Bingley, UK: Emerald Group Publishing Limited. https://doi.org/10.1108/S1057-6290(2012)0000014008

Hansen, H.B., Siegel, C.E., Case, B.G., Bertollo, D.N., DiRocco, D., & Galanter, M. (2013). Variation in use of buprenorphine and methadone treatment by racial, ethnic, and income characteristics of residential social areas in New York City. *The Journal of Behavioral Health Services & Research*, 40(3), 367–377. https://doi.org/10.1007/s11414-013-9341-3

Harm Reduction International. (2020). What is harm reduction? Retrieved from https://www.hri.global/what-is-harm-reduction

Hassan, A. (2019). Ohio doctor charged with killing 25 patients in fentanyl overdoses. *The New York Times*. https://www.nytimes.com/2019/06/05/us/ohio-doctor-murder-fentanyl-overdose.html

Hayden, N. (2017, April 14). "It's everywhere": Heroin addiction doesn't discriminate. *The Times Herald*. https://www.thetimesherald.com/story/news/local/2017/04/14/face-heroin-addict/100421414/

Hearing of the Committee on Health, Education, Labor, and Pensions. United States Senate. (2022). 107th Congress, 2nd session (pp. 13–18). Washington, DC: U.S. Government Printing Office. https://www.gpo.gov/fdsys/pkg/CHRG-107shrg77770/pdf/CHRG-107shrg77770.pdf

Hearn, A. (2010). Structuring feeling: Web 2.0, online ranking and rating, and the digital "reputation" economy. *Ephemera*, 10(3/4), 421–438.

Hedegaard, H., Warner, M., Miniño, A.M. (2017). *Drug overdose deaths in the United States, 1999–2016.* NCHS Data Brief, no 294. Hyattsville, MD: National Center for Health Statistics.

Helm, K.A. (2007). Protecting public health from outside the physician's office: A century of FDA regulation from drug safety labeling to off-label drug promotion. *Fordham Intellectual Property, Media and Entertainment Law Journal*, 18(1), 117–187.

Herzberg, David. (2020). *White market drugs: Big pharma and the hidden history of addiction in America.* Chicago: University of Chicago Press.

Hilts, P.J. (2003). *Protecting America's health: The FDA, business, and one hundred years of regulation.* Chapel Hill: The University of North Carolina Press.

Hirschfeld Davis, J. (2017, October 26). Trump declares opioid crisis a "health emergency" but requests no funds. *The New York Times*. https://www.nytimes.com/2017/10/26/us/politics/trump-opioid-crisis.html

Hochschild, A.R. (2018). *Strangers in their own land: Anger and mourning on the American right.* New York: The New Press.

Hoffman, K.M., Trawalter, S., Axt, J.R., & Oliver, M.N. (2016). Racial bias in pain assessment and treatment recommendations, and false beliefs about biological differences

between blacks and whites. *Proceedings of the National Academy of Sciences USA, 113*(16), 4296–4301.
Hoffmann, N.G., Olofsson, O., Salen, B., & Wickstrom, L. (1995). Prevalence of abuse and dependency in chronic pain patients. *International Journal of the Addictions, 30*(8), 919–927.
Homeland Security and Government Affairs Committee (HSGAC). (2018). *Fueling an epidemic—Report Two: Exposing financial ties between opioid manufacturers and third party advocacy groups.* Washington, DC: U.S. Senate. https://www.hsdl.org/?view&did=808171
Home page. (2004, April 2). *Partners Against Pain.* Retrieved from https://web.archive.org/web/20040402052021/http://www.partnersagainstpain.com/index.aspx?sid=27
Home page. (2015, October 11). *Team Against Opioid Abuse.* Retrieved from https://web.archive.org/web/20151011032341/http://teamagainstopioidabuse.com:80/
Human Rights Watch. (2021, August 24). More than 140 groups urge DOJ to end over-criminalization of fentanyl-related substances. https://www.hrw.org/news/2021/08/24/more-140-groups-urge-doj-end-over-criminalization-fentanyl-related-substances#_ftnref10
IASP. (1994). IASP taxonomy. Retrieved from http://www.iasp-pain.org/Taxonomy
Illouz, E. (2008). *Saving the modern soul: Therapy, emotions, and the culture of self-help.* Berkeley, CA: University of California Press.
Ingraham, C., & Johnson, C.Y. (2016). Ohio city shares shocking photos of adults who overdosed with a small child in the car. *The Washington Post.* https://www.washingtonpost.com/news/wonk/wp/2016/09/09/ohio-city-shares-shocking-photos-of-adults-who-overdosed-with-a-small-child-in-their-car/
Irvine, M.A., Kuo, M., Buxton, J.A., Balshaw, R., Otterstatter, M., Macdougall, L., Milloy, M-J., Bharmal, A, Henry, B., Tyndall, M., Coombs, D., & Gilbert, M. (2019). Modelling the combined impact of interventions in averting deaths during a synthetic-opioid overdose epidemic. *Addiction. 114*(9), 1602–1613.
Jacox, A.K. (1979). Assessing pain. *American Journal of Nursing, 79*(5), 895–900.
Jacox A., Carr, D.B., Payne, R. (1994). *Clinical practice guideline: Management of cancer pain* (AHCPR publication 94-0592). Rockville, MD: U.S. Department of Health and Human Services, Agency for Health Care Policy and Research.
Jacques, S., & Allen, A. (2015). Drug market violence: Virtual anarchy, police pressure, predation and retaliation. *Criminal Justice Review, 40*(1), 87–99.
Jacques, S., & Wright, R. (2015). *Code of the suburb: Inside the world of young middle-class drug dealers.* Chicago: University of Chicago Press.
Jaffe, J.H. (1980). Drug addiction and drug abuse. In A.G. Gilman, L.S. Goodman, & B.A. Gilman (Eds.), *The pharmacological basis of therapeutics* (8th ed.) (pp. 522–573). New York: Macmillan.
Jaffe, J.H. (1989). Misinformation: Euphoria and addiction. In C.S. Hill and W.S. Fields (Eds.), *Advances in pain research and therapy* (pp. 163–174). New York: Raven Press.
Jasanoff, S. (1990). *The fifth branch: Science advisors as policymakers.* Cambridge, MA: Harvard University Press.
Jenkins, J. (2002, February 2). Prepared statement. In *OxyContin: Balancing risks and benefits.* Retrieved from https://www.govinfo.gov/app/details/CHRG-107shrg77770/summary
jman1982. (2012, April 19). Use a dremel for the new Opana ER's [Online forum comment]. Message posted to https://drugs-forum.com/threads/use-a-dremel-for-the-new-opana-ers.182993/

Joint Commission on the Accreditation of Health Care Organizations (JCAHO). (2001). Pain standards for 2001. https://www.jointcommission.org/assets/1/6/2001_Pain_St andards.pdf

Jones, C.M. (2013). Heroin use and heroin use risk behaviors among nonmedical users of prescription opioid pain relievers - United States, 2002–2004 and 2008–2010. *Drug and Alcohol Dependence, 132,* 95–100.

Jones, J.D., Mogali, S. & Comer, S.D. (2012). Polydrug abuse: A review of opioid and benzodiazepine combination use. *Drug and Alcohol Dependence, 125*(1–2), 8–18.

Joranson, D.E., Carrow, G.M., Ryan, K.M., Schaefer, L., Gilson, A.M., Good, P., Eadie, J., Peine, S., & Dahl, J.L. (2002). Pain management and prescription monitoring. *Journal of Pain and Symptom Management, 23*(3), 231–238.

Joranson, D.E., & Gilson, A.M. (1994). Policy issues and imperatives in the use of opioids to treat pain in substance abusers. *Medicine and Ethics, 22*(3), 215–223.

Kaplan, K. (2018, March 29). Opioid overdose deaths are still rising in nearly every segment of the country, CDC says. *Los Angeles Times.* http://www.latimes.com/science/sci encenow/la-sci-sn-opioid-overdose-deaths-20180329-htmlstory.html

Katz, J. (2017, June 5). Drug deaths in America are rising faster than ever. *The New York Times.* https://www.nytimes.com/interactive/2017/06/05/upshot/opioid-epidemic-drug-overdose-deaths-are-rising-faster-than-ever.html

Katz, J., & Goodnough, A. (2017, December 22). The opioid crisis is getting worse, particularly for black Americans. *The New York Times.* https://www.nytimes.com/interact ive/2017/12/22/upshot/opioid-deaths-are-spreading-rapidly-into-black-america.html

Keier, M. (1988). Dope fiends and degenerates: The gendering of addiction in the early twentieth century. *Journal of Social History, 31*(4), 809–822.

Keilman, J. (2017, June 5). Chronic pain patients say opioid crackdown is hurting them. *Chicago Tribune.* http://www.chicagotribune.com/lifestyles/health/ct-opioid-patients-backlash-met-20170603-story.html

Kertesz, S.G., & Gordon, A.J. (2017, February 24). Strict limits on opioid prescribing risk the "inhumane treatment" of pain patients. *STAT News.* www.statnews.com/2017/02/24/opioids-prescribing-limits-pain-patients/

Kertesz, S.G., & Satel, S. (2017, August 17). Some people still need opioids. *Slate.* http://www.slate.com/articles/health_and_science/medical_examiner/2017/08/cutting_down_on_opioids_has_made_life_miserable_for_chronic_pain_patients.html

Kessler, D.A. (2017, May 6). The opioid epidemic we failed to foresee. *The New York Times.* https://www.nytimes.com/2016/05/07/opinion/the-opioid-epidemic-we-failed-to-foresee.html

Klinenberg, E. (2002). *Heat wave: A social autopsy of disaster in Chicago.* Chicago: University of Chicago Press.

Knight, K. (2015). *Addicted.pregnant.poor.* Durham, NC: Duke University Press.

Knight, K., Kushel, M., Chang, J.S., Zamora, K., Caesar, R., Hurstak, E., & Miaskowski, C. (2017). Opioid pharmacovigilance: A clinical-social history of the changes in opioid prescribing for patients with co-occurring chronic non-cancer pain and substance use. *Social Science and Medicine, 186,* 87–95.

Kolb, L. (1927). Clinical contribution to drug addiction: The struggle for cure and the conscious reasons for relapse. *Journal of Nervous and Mental Disease, 66*(1), 22–45.

Kolb, L. (1928). Drug addiction: A study of some medical cases. *Archives of Neurology & Psychiatry, 20*(1), 171. https://doi.org/10.1001/archneurpsyc.1928.02210130174012

Kouyanou, K., Pither, C.E., & Wessely, S. (1997). Medication misuse, abuse and dependence in chronic pain patients. *Journal of Psychosomatic Research, 43*(5), 497–504.

Lakoff, A. (2007). The right patients for the drug: Managing the placebo effect in antidepressant trials. *BioSocieties, 2*, 57–71.

Lakoff, A. (2017). *Unprepared: Global health in a time of emergency*. Oakland, CA: University of California Press.

Lakoff, A., & Collier, S.J. (Eds.). (2008). *Biosecurity interventions: Global health & security in question*. New York: Columbia University Press.

Langston, N. (2010). *Toxic bodies: Hormone disruptors and the legacy of DES*. New Haven, CT: Yale University Press.

Lassiter, M.D. (2015). Impossible criminals: The suburban imperatives of America's war on drugs. *Journal of American History, 102*(1), 126–140. https://doi.org/10.1093/jahist/jav243

Lavitt, J. (2015, November 13). Interview with Michael Botticelli, Obama's "Recovery Czar." *The Fix*. https://www.thefix.com/obama-drug-czar-michael-botticelli-now-recovery-czar

Leff, D.N. (1976). Management of chronic pain: Medicine's new growth industry. *Medical World News, 18*, 54–77.

LeFort, Sandra M., Gray-Donald, Katherine, Rowat, Katherine M., & Jeans, Mary Ellen. (1998). Randomized controlled trial of a community-based psychoeducation program for the self-management of chronic pain *Pain, 74*(2), 297–306. https://doi.org/10.1016/S0304-3959(97)00190-5

Levenson, E., & del Valle, L. (2020). Judge clears path for Philadelphia nonprofit to open safe-injection site to combat overdoses. *CNN*. https://www.cnn.com/2020/02/26/us/philadelphia-supervised-injection-site/index.html

Lorig, K.R., & Holman, H.R. (2003). Self-management education: History, definition, outcomes and mechanisms. *Annals of Behavioral Medicine, 26*, 1–7. (Note: Provides a comprehensive overview of self-management.) https://tinyurl.com/wysusf8p

Lyons Lavey Nickel Swift, Inc. (1997). *"I got my life back": Patients in pain tell their story*. [VHS]. New York, NY: Purdue Pharma L.P.

Lyons Lavey Nickel Swift, Inc. (1999). *"I got my life back" part 2: A two-year follow up of patients in pain*. [VHS]. New York, NY: Purdue Pharma L.P.

Mann, C.C., & Plummer, M.L. (1991). *The aspirin wars: Money, medicine, and 100 years of rampant competition*. New York: Alfred A. Knopf.

Marcus, D.A. (2000). Treatment of nonmalignant chronic pain. *American Family Physician, 61*(5), 1331–1338.

Marez, C. (2004). *Drug wars: The political economy of narcotics*. Minneapolis: University of Minnesota Press.

Mariani, M. (2015, March 4). How the American opiate epidemic was started by one pharmaceutical company. *The Week*. http://theweek.com/articles/541564/how-american-opiate-epidemic-started-by-pharmaceutical-company

Markey, E.J. (2016, February). "Abuse-Deterrent Opioid" is an oxymoron: FDA's "Action Plan" preserves its "No Help Wanted" status quo. *Office of Senator Edward J. Markey* (D-MA). https://www.markey.senate.gov/imo/media/doc/2016-02-19-Markey-ADF-Opioid-timeline.pdf

Marks, J.H. (2020). Lessons from corporate influence in the opioid epidemic: Toward a norm of separation. *Journal of Bioethical Inquiry, 17*(2), 173–189.

Maron, D.F. (2018, March 20). Is Trump's opioid strategy a "War on Drugs" relapse? *Scientific American*. https://www.scientificamerican.com/article/is-trumps-opioid-strategy-a-war-on-drugs-relapse/

McGee, M. (2005). *Self-help, Inc: Makeover culture in American life*. Oxford: Oxford University Press.

McGoey, L. (2019). *The unknowers: How strategic ignorance rules the world*. Zed Books.

Mehlsen, Mimi, Hegaard, Lea, Ørnbøl, Eva, Jensen, Jens Søndergaard, Fink, Per, & Frostholm, Lisbeth. (2017). The effect of a lay-led, group-based self-management program for patients with chronic pain: A randomized controlled trial of the Danish version of the Chronic Pain Self-Management Programme. *PAIN*, *158*(8), 1437–1445. 10.1097/j.pain.0000000000000931

Meier, B. (2001, October 28.) Overdoses of painkiller are linked to 282 deaths. *The New York Times*. https://www.nytimes.com/2001/10/28/us/overdoses-of-painkiller-are-linked-to-282-deaths.html

Meier, B. (2002, April 15). OxyContin deaths may top early count. *The New York Times*. http://www.nytimes.com/2002/04/15/us/oxycontin-deaths-may-top-early-count.html

Meier, B. (2007, May 10). In guilty plea, OxyContin maker to pay $600 million. *The New York Times*. http://www.nytimes.com/2007/05/10/business/11drug-web.html

Meier, B. (2013, June 22). Profiting from pain. *The New York Times*. http://www.nytimes.com/2013/06/23/sunday-review/profiting-from-pain.html

Meldrum, M.L. (2003). A capsule history of pain management. *JAMA*, *290*(18), 2470. https://doi.org/10.1001/jama.290.18.2470

Meldrum, M. (2005). The ladder and the clock: Cancer pain and public policy at the end of the twentieth century. *Journal of Pain and Symptom Management*, *29*(1), 41–54.

Meldrum, M. (2007). A brief history of multidisciplinary management of chronic pain, 1900– 2000. In M.E. Schatman, A. Campbell, & J.D. Loeser (Eds.), *Chronic pain management: Guidelines for multidisciplinary program development* (pp. 1–13). New York, NY: Informa Healthcare USA Inc.

Meldrum, M. (2016). The ongoing opioid prescription epidemic: Historical context. *American Journal of Public Health*, *106*(8), 1365–1366. https://doi.org/10.2105/AJPH.2016.303297

Melzack, R., & Wall, P. (1965). Pain mechanisms: A new theory. *Science*, *150*(3699), 971–979.

Merica, D. (2016, October 26). Trump declares opioid epidemic a national public health emergency. *CNN*. http://www.cnn.com/2017/10/26/politics/donald-trump-opioid-epidemic/index.html

Merica, D. (2018). Trump pushes death penalty for some drug dealers. *CNN*. https://edition.cnn.com/2018/03/19/politics/opioid-policy-trump-new-hampshire/index.html

Meyer, R.J. (2005). FDA's Role in Preventing Prescription Drug Abuse, Statement before the House Committee on Government Reform (September 13). FDA.

Miller, T. (2008). *Makeover nation: The United States of reinvention*. Columbus: Ohio State University Press.

Moghe, S. (2016, October 14). Opioid history: From "wonder drug" to abuse epidemic. *CNN*. http://edition.cnn.com/2016/05/12/health/opioid-addiction-history/index.html

Moor, E. (2003). Branded spaces: The scope of new marketing. *Journal of Consumer Culture*, *3*(1), 39–60.

Moor, L. (2007). *The rise of brands*. Oxford; New York: Berg.

Moshtaghian, A., & Meilhan, P. (2018, August 29). Prince family files wrongful death lawsuit. *CNN*. https://www.cnn.com/2018/08/29/health/prince-family-civil-lawsuit/index.html

Moynihan, R., & Cassels, A. (2005). *Selling sickness: How the world's biggest pharmaceutical companies are turning us all into patients.* Vancouver: Greystone Books.

Muhuri, P., Gfroerer, J., & Davies, M.C. (2013). Associations of nonmedical pain reliever use and initiation of heroin use in the United States. [Research report, August 2013] *CBHSQ Data Review.* https://www.samhsa.gov/data/sites/default/files/DR006/DR006/nonmedical-pain-reliever-use-2013.htm

Mukherjee, R., & Banet-Weiser, S. (2012). *Commodity activism: Cultural resistance in neoliberal times.* New York: New York University Press.

Murphy, K.R., Han, J.L., Yang, S., Hussaini, S.M.Q., Elsamadicy, A.A., Parente, B., Xie, J., Pagadala, P., & Lad, S.P. (2017). Prevalence of specific types of pain diagnoses in a sample of United States adults. *Pain Physician, 20*, E257–E268.

National Cancer Institute. (2013). 61-year trends in U.S. cancer death rates. https://seer.cancer.gov/archive/csr/1975_2010/results_merged/topic_historical_mort_trends.pdf

NCADD. (2015, March 15). Reformulated version of OxyContin can still be abused. https://www.ncadd.org/blogs/in-the-news/entry/reformulated-version-of-oxycontin-can-still-be-abused

NCADD. (2016, May 5). 44% of Americans know someone who has been addicted to prescription painkillers. https://www.ncadd.org/blogs/in-the-news/44-of-americans-know-someone-who-has-been-addicted-to-prescription-painkillers

Nestler, E.J. (2004). Molecular mechanisms of drug addiction. *Neuropharmacology, 47*, 24–32. https://doi.org/10.1016/j.neuropharm.2004.06.031

Netherland, J., & Hansen, H.B. (2016). The war on drugs that wasn't: Wasted whiteness, "dirty doctors," and race in media coverage of prescription opioid misuse. *Culture, Medicine, and Psychiatry, 40*(4), 664–686. https://doi.org/10.1007/s11013-016-9496-5

Netherland, J., & Hansen, H. (2017). White opioids: Pharmaceutical race and the war on drugs that wasn't. *BioSocieties, 12*(2), 217–238. https://doi.org/10.1057/biosoc.2015.46

Ng, J., Sutherland, C., & Kolber, M.R. (2017). Does evidence support supervised injection sites? *Canadian Family Physician (Medecin de famille canadien), 63*(11), 866.

NIDA. (2015). Overdose death rates. Retrieved from https://www.drugabuse.gov/related-topics/trends-statistics/overdose-death-rates

NIDA. (2017). Overdose death rates. Retrieved from https://www.drugabuse.gov/related-topics/trends-statistics/overdose-death-rates

NIDA. (2020). Opioid overdose crisis. Retrieved from https://www.drugabuse.gov/drug-topics/opioids/opioid-overdose-crisis

NIDA. (2022). Overdose death rates. Retrieved from https://nida.nih.gov/drug-topics/trends-statistics/overdose-death-rates

NIH. (2007). The science of addiction: Drugs, brains, and behavior. *NIH Medline Plus, 2*(2, Spring). https://medlineplus.gov/magazine/issues/spring07/articles/spring07pg14-17.html

Noble, B., Clark, D., Meldrum, M., ten Have, H., Seymour, J., Winslow, M., & Paz, S. (2005). The measurement of pain, 1945–2000. *Journal of Pain and Symptom Management, 29*(1), 14–21.

Norton, G.R., Asmundson, G.J.G., Norton, R.G., & Craig, K.D. (1999). Growing pain: 10-year research trends in the study of chronic pain and headache. *Pain, 79*, 59–65.

Offit, P.A. (2017, April 1). The doctors who started the opioid epidemic. *The Daily Beast.* http://www.thedailybeast.com/the-doctors-who-started-the-opioid-epidemic

Ohio Department of Health. (2017). 2016 Ohio drug overdose data: General findings. Retrieved from http://www.odh.ohio.gov/-/media/ODH/ASSETS/Files/health/injury-prevention/2016-Ohio-Drug-Overdose-Report-FINAL.pdf

ONDCP. (2021). The Biden-Harris administration's statement of drug policy priorities for year one. Executive Office of the President. https://www.whitehouse.gov/wp-content/uploads/2021/03/BidenHarris-Statement-of-Drug-Policy-Priorities-April-1.pdf?fbclid=IwAR2TBk34U_XRqlqK_pAYnUd_9f7zY3IbCQI9KxI6S5eYeRJdFzl9B09hZ84

Oral History Interview with John J. Bonica, March 9–12, (1993a). (Ms. Coll. no. 127.7), John C. Liebeskind History of Pain Collection, History and Special Collections Division, Louise M. Darling Biomedical Library, University of California, Los Angeles.

Oral History Interview with Wilbert E. Fordyce, July10, (1993b). (Ms. Coll. no. 127.1), John C. Liebeskind History of Pain Collection, History and Special Collections Division, Louise M. Darling Biomedical Library, University of California, Los Angeles.

Ordway, R. (2000, April 6). Narcotic abuse on the rise: Pharmaceutical drug fraud, misuse worries officials. *Bangor Daily News*, 4.

Ornstein, C., & Weber, T. (2012, May 8). American Pain Foundation shuts down as senators launch investigation of prescription narcotics. *ProPublica.* https://www.propublica.org/article/senate-panel-investigates-drug-company-ties-to-pain-groups

Ortner, N. (2014). *Tapping solution: A revolutionary system for stress-free living.* Carlsbad, CA: Hay House, Inc.

Ortner, N. (Director), & Polizzi, N. (Director). (2008). *The tapping solution* [Film]. Newtown, CT: The Tapping Solution, L.L.C.

Ouellette, L. (2012). Citizen brand: ABC and the do good turn in US television. In R. Mukherjee & S. Banet-Weiser (Eds.), *Commodity activism: Cultural resistance in neoliberal times* (pp. 57–75). New York: New York University Press.

OxyContin: Balancing risks and benefits: Hearings before the Committee on Health, Education, Labor, and Pensions. (2002). Senate, 107th Congress. https://www.govinfo.gov/content/pkg/CHRG-107shrg77770/html/CHRG-107shrg77770.htm

Pasero, C.L., & McAffery, M.M. (1997). Pain ratings: The fifth vital sign. *American Journal of Nursing, 97*(2), 15.

Paulozzi, L.J., Mack, K.A., & Hockenberry, J.M. (2014). *Vital signs: Variation among states in prescribing pain relievers and benzodiazepines—United States.* Atlanta: Center for Disease Control and Prevention.

Pazanowski, M.A. (2018, July 17). Lawsuits targeting doctors next opioid litigation wave? *Bloomberg Law.* https://news.bloomberglaw.com/us-law-week/lawsuits-targeting-doctors-next-opioid-litigation-wave

Pew Research Center. (2014a, April 2). America's new drug policy landscape: Two-thirds favor treatment, not jail, for use of heroin, cocaine. http://www.people-press.org/2014/04/02/americas-new-drug-policy-landscape/

Pew Research Center. (2014b, July 18). Chart of the week: The black-white gap in incarceration rates. http://www.pewresearch.org/fact-tank/2014/07/18/chart-of-the-week-the-black-white-gap-in-incarceration-rates/

Popper, N. (2017, June 10). Opioid dealers embrace the dark web to send deadly drugs by mail. *The New York Times.* https://www.nytimes.com/2017/06/10/business/dealbook/opioid-dark-web-drug-overdose.html

Portenoy, R.K. (1996). Report from the International Association for the Study of Pain taskforce. *Journal of Pain and Symptom Management, 12*(2), 93–96.

Portenoy, R.K., & Foley, K.M. (1986). Chronic use of opioid analgesics in non-malignant pain: Report of 38 cases. *Pain, 25*(2), 171–186.

Poston, B. (2018, February 10). Capping years of criticism, Purdue Pharma will stop promoting its opioid drugs to doctors. *Los Angeles Times.* http://www.latimes.com/local/lanow/la-me-ln-purdue-marketing-20180210-story.html

Purdue Pharma. (1995, December). *New drug application for OxyContin.* NDA 20-553. Purdue Pharma: Stamford, CT.

Purdue Press Release. (2015, August 17). Purdue Pharma L.P. launches TeamAgainstOpioidAbuse.com. https://www.prnewswire.com/news-releases/purdue-pharma-lp-launches-teamagainstopioidabusecom-300129051.html

Radden Keefe, P. (2017, October 30). The family that built an empire of pain. *The New Yorker.* https://www.newyorker.com/magazine/2017/10/30/the-family-that-built-an-empire-of-pain

Raikhel, E.A., and Garriott, W.C. (Eds.). (2013). *Addiction trajectories.* Durham and London: Duke University Press.

Reuters. (2010, April 5). Harder-to-break OxyContin pill wins approval. *The New York Times.* http://www.nytimes.com/2010/04/06/business/06oxy.html

Rey, R. (1995). *The history of pain.* Cambridge, MA: Harvard University Press.

Rodgers, M.C. (1991). Subjective pain testimony in disability determination proceedings: Can pain alone be disabling? *California Western Law Review, 28*(1), 173–212.

Rose, N. (1999). *Powers of freedom: Reframing political thought.* Cambridge, UK: Cambridge University Press.

Rose, N. (2006). *The politics of life itself: Biomedicine, power, and subjectivity in the twenty-first century.* Princeton, NJ: Princeton University Press.

Rowland, C. (2004, January 18). FDA's economist in chief. *The Boston Globe.* http://archive.boston.com/business/healthcare/articles/2004/01/18/fdas_economist_in_chief/

Rudd, R.A., Seth, P., David, F., & Scholl, L. (2016). Increases in drug and opioid-involved overdose deaths - United States, 2010–2015. *Morbidity and Mortality Weekly Report, 65,* 1445–1452. http://dx.doi.org/10.15585/mmwr.mm655051e1

Ryan, H., Girion, L., & Glover, S. (2016, May 5). "You want a description of hell"? OxyContin's 12-hour problem. *Los Angeles Times.* http://www.latimes.com/projects/oxycontin-part1/

SAMHSA. (2014). Behavioral health trends in the United States: Results from the 2014 National Survey on Drug Use and Health. https://www.samhsa.gov/data/sites/default/files/NSDUH-FRR1-2014/NSDUH-FRR1-2014.pdf

SAMHSA. (2017, February 9). Apply to increase patient limits. Retrieved from https://www.samhsa.gov/medication-assisted-treatment/buprenorphine-waiver-management/increase-patient-limits

SAMHSA. (2018). Facilities providing all medication assisted treatments and accepting Medicaid. Retrieved from http://opioid.amfar.org/indicator/Med_AMAT_fac

SAMHSA. (2020). Key substance use and mental health indicators in the United States: Results from the 2020 National Survey on Drug Use and Health. https://www.samhsa.gov/data/sites/default/files/reports/rpt35325/NSDUHFFRPDFWHTMLFiles2020/2020NSDUHFFR1PDFW102121.pdf

Satterfield, J. (2017, June 13). Tennessee counties sue drug makers over opioid epidemic. *USA Today.* https://www.usatoday.com/story/news/nation-now/2017/06/14/tennes see-counties-sue-drugmakers-over-opioid-epidemic/394968001/

Saunders, C. (1959, October 23). Care of the dying: Control of pain in terminal cancer. *Nursing Times,* 1031–1032.

Saunders, C. (1978). *The management of terminal malignant disease* (1st ed.). London: Edward Arnold.

Schuster, Janice L. (1999, February 2). Addressing patients' pain. *The Washington Post.* https://www.washingtonpost.com/archive/lifestyle/wellness/1999/02/02/addressing-patients-pain/f974e07e-ce12-4b5b-977a-5a63b433f82f/

Schwartz, Y. (2012, June 19). Painkiller use breeds new face of heroin addiction. *NBC News.* http://www.hcdrugfree.org/drug-alcohol-news/2015/2/13/painkiller-use-bre eds-new-face-of-heroin-addiction

Seelye, K.Q. (2015, October 30). In heroin crisis, white families seek gentler war on drugs. *The New York Times.* https://www.nytimes.com/2015/10/31/us/heroin-war-on-drugs-parents.html.

Sessler, N.E., Downing, J.M., Kale, H., Chilcoat, H.D., Baumgartner, T.F., & Coplan, P.M. (2014). Reductions in reported deaths following the introduction of extended-release oxycodone (OxyContin) with an abuse-deterrent formulation. *Pharmacoepidemiology Drug Safety, 23*(12), 1238–1246.

Severtson, S.G., Bartelson, B.B., Davis, J.M., Muñoz, A., Schneider, M.F., Chilcoat, H., Coplan, P.M., Surratt, H., & Dart, R.C. (2013). Reduced abuse, therapeutic errors, and diversion following reformulation of extended-release oxycodone in 2010. *Journal of Pain, 14*(10), 1122–1130.

Seymour, J., Clark, D., & Winslow, M. (2005). Pain and palliative care: The emergence of new specialties. *Journal of Pain and Symptom Management, 29*(1), 2–13.

Shah, N. (2001). *Contagious divides: Epidemics and race in San Francisco's Chinatown.* Berkeley: University of California Press.

Siegel, E. (2018, March 20). Opioid epidemic so dangerous, says CDC, it's finally killing as many Americans as guns. *Forbes.* https://www.forbes.com/sites/startswithabang/2018/03/20/opioid-epidemic-so-dangerous-says-cdc-its-finally-killing-as-many-americ ans-as-guns/#6e26752e6c21

Silverman, R. (2014, November 17). NFL runs on piles of painkillers. *The Daily Beast.* http://www.thedailybeast.com/articles/2014/11/17/the-nfl-runs-on-piles-of-painkill ers.html

Skocpol, T. (1995). *Protecting soldiers and mothers: The political origins of social policy in the United States.* Cambridge: Harvard University Press.

Smith, C. (2018, August 21). News: 27 states have filed lawsuits against opioid manufacturer. *Addiction Center.* https://www.addictioncenter.com/community/states-filed-suits-against-opioid/

Social Security Disability Benefits Reform Act of 1984. (1984). Pub. L. No. 98-460, 98 Stat. 1794, codified as 42 U.S.C. §§1799–1800.

Spencer, M.R., Warner, M., & Bastion, B.A. (2019). Drug overdose deaths involving fentanyl, 2011–2016. *National Vital Statistics Report, 68*(3). U.S. Department of Health and Human Services. https://www.cdc.gov/nchs/data/nvsr/nvsr68/nvsr68_03-508.pdf

Stanford School of Medicine Palliative Care. (2017). Opioid conversion. Retrieved from https://dx.stanford.edu/resources/OpioidConversion2.html

Stanforth, E., Kostiuk, M., & Garrison, P. (2016). Correlates of engaging in drug distribution in a national sample. *Psychology of Addictive Behaviors, 30*(1), 141.

Stanley, T.H. (1992, April). The history and development of the fentanyl series. *Journal of Pain and Symptom Management, 7*(3), S3–S7. https://doi.org/10.1016/0885-3924(92)90047-L

Sullivan, S.P. (2017, March 2). N.J. doctor charged in opioid overdose death of patient. *New Jersey Real-Time News.* http://www.nj.com/news/index.ssf/2017/03/nj_doctor_indicted_over_claims_he_prescribed_opioi.html

Sunstein, C.R. (2002). *Risk and reason: Safety, law, and the environment.* Cambridge; New York: Cambridge University Press.

Swenson, K. (2010). *Lifestyle drugs and the neoliberal family.* New York: Peter Lang.

Temple, J. (2016). *American pain: How a young felon and his ring of doctors unleashed America's deadliest drug epidemic.* Guilford, CT: Lyons Press.

Timeline of selected FDA activities and significant events addressing opioid misuse and abuse. (n.d.). Retrieved from https://www.fda.gov/drugs/information-drug-class/timeline-selected-fda-activities-and-significant-events-addressing-substance-use-and-overdose

Timmermans, S., & Berg, M. (2003). *The gold standard: The challenge of evidence-based medicine and standardization in health care.* Philadelphia: Temple University Press.

Tough, P. (2001, July 29). The alchemy of OxyContin. *The New York Times.* http://www.nytimes.com/2001/07/29/magazine/the-alchemy-of-oxycontin.html

Tousignant, N.R. (2006). *Pain and the pursuit of objectivity: Pain-measuring technologies in the United States, c. 1890–1975.* [Dissertation, McGill University]. Retrieved from https://escholarship.mcgill.ca/concern/theses/tm70n0508

Treede, R-D., Rief, W., Barke, A., Aziz, Q., Bennett, M.I., Benoliel, R., Cohen, M., Evers, S., Finnerup, N.B., First, M.B., Giamberardino, M.A., Kaasa, S., Kosek, E., Lavand'homme, P., Nicholas, M., Perrot, S., Scholz, J., Schug, S., Smith, B.H., . . . Wang, S.J. (2015). A classification of chronic pain for ICD-11. *Pain, 156*(6), 1003–1007.

Trost, Z., Sturgeon, J., Guck, A., Ziadni, M., Nowlin, K., Goodin, B., & Scott, W. (2019). Examining injustice appraisals in a racially diverse sample of individuals with chronic low back pain. *The Journal of Pain, 20*(1), 83–96.

Turner, J.A., Calsyn, D.A., Fordyce, W.E., & Ready, L.B. (1982). Drug utilization patterns in chronic pain patients. *Pain, 12,* 357–363.

United States Courts. (2017a, July 25). Mandatory minimum sentences decline, sentencing commission says. http://www.uscourts.gov/news/2017/07/25/mandatory-minimum-sentences-decline-sentencing-commission-says

United States Courts. (2017b, June 26). Mandatory minimum terms result in harsh sentencing. https://web.archive.org/web/20101208070157/http://www.uscourts.gov/News/NewsView/07-06-26/Mandatory_Minimum_Terms_Result_In_Harsh_Sentencing.aspx

United States Government Printing Office. (2002). *OxyContin: Balancing risks and benefits. Hearing of the Committee on Health, Education, Labor, and Pensions.* United States Senate, 107th Congress, 2nd session. Washington, DC: U.S. Government Printing Office. https://www.govinfo.gov/content/pkg/CHRG-107shrg77770/html/CHRG-107shrg77770.htm

United States House Committee on Government Reform. (2005, September 13). *FDA's role in preventing prescription drug abuse.* [Statement of Robert J. Meyer, MD, Director, Office of Drug Evaluation II, Center for Drug Evaluation and Research, Food and

Drug Administration. House Committee on Government Reform, US House of Representatives].
United States Sentencing Commission. (2021). *Fentanyl and fentanyl analogues: Federal trends and trafficking patterns.* https://www.ussc.gov/sites/default/files/pdf/research-and-publications/research-publications/2021/20210125_Fentanyl-Report.pdf
U.S. Department of Health and Human Services (HHS) Office of the Surgeon General. (2018). *U.S. Surgeon General's advisory on naloxone and opioid overdose* [Internet]. Washington DC: HHS. Retrieved from https://www.hhs.gov/surgeongeneral/repo rts-and-publications/addiction-and-substance-misuse/advisory-on-naloxone/ index.html
U.S. Department of Justice. (2018, July 12). Attorney General Jeff Sessions announces the formation of Operation Synthetic Opioid Surge (S.O.S.). https://www.justice.gov/ opa/pr/attorney-general-jeff-sessions-announces-formation-operation-synthetic-opi oid-surge-sos
USPS. (2019). Statement of Gary R. Barksdale, Chief Postal Inspector, United States Postal Inspection Service, before the Committee on Energy and Commerce Subcommittee on Oversight and Investigations United States House of Representatives. https://about.usps.com/news/testimony/2019/pr19_cpi0716.htm
Valverde, M. (1998). *Diseases of the will: Alcohol and the dilemmas of freedom.* Cambridge: Cambridge University Press.
Van Zee, A. (2009). The promotion and marketing of OxyContin: Commercial triumph, public health tragedy. *American Journal of Public Health, 99*(2), 221–227.
Vardanyan, R.S., & Hruby, V.J. (2014). Fentanyl-related compounds and derivatives: Current status and future prospects for pharmaceutical applications. *Future Medicinal Chemistry, 6*(4), 385–412. https://www.ncbi.nlm.nih.gov/pmc/artic les/PMC4137794/
Vijayaraghavan M., Penko J., Bangsberg, D.R., Miaskowski, C., & Kushel, M.B. (2013). Opioid analgesic misuse in a community-based cohort of HIV-infected indigent adults. *JAMA Internal Medicine, 173*(3), 235–7.
Von Drehle, D. (2018, March 2). The opioid crisis is a government failure of epic proportions. *The Washington Post.* https://www.washingtonpost.com/opinions/the-opioid-crisis-is-a-government-failure-of-epic-proportions/2018/03/02/5c1decf6-1e4f-11e8-b2d9-08e748f892c0_story.html?utm_term=.aa0ab055b30b
Vowles, K.E., McEntee, M.L., Julnes, P.S., Frohe, T., Ney, J.P., & van der Goes, D.N. (2015). Rates of opioid misuse, abuse, and addiction in chronic pain: A systematic review and data synthesis. *Pain, 156*(4), 569–576.
Vrecko, S. (2010). Birth of a brain disease: Science, the state and addiction neuropolitics. *History of the Human Sciences, 23*(4), 52–67.
Wagner, J., Bernstein, L., & Johnson, J. (2017, October 26). Trump declares opioid crisis a public health emergency; crisis say plan falls short. *The Washington Post.* https://www.washingtonpost.com/politics/trump-declares-opioid-crisis-a-public-health-emerge ncy-critics-say-plan-falls-short/2017/10/26/8883762e-ba60-11e7-be94-fabb0f1e9ffb_ story.html?utm_term=.186772e0e677
Wagner, J., & Zezima, K. (2018, March 19). In N.H., Trump pledges to "get tough" on drug crime, unveils opioid epidemic plan. *The Washington Post.* https://www.washingtonp ost.com/politics/trump-pledges-to-get-tough-on-drug-crime-unveils-opioid-epide mic-plan-in-new-hampshire/2018/03/19/908585a2-2b8e-11e8-8688-e053ba58f1e4_ story.html?utm_term=.876f8715c9bd

Wailoo, K. (2015). *Pain: A political history*. Baltimore: Johns Hopkins University Press.

Wall, P.D. (1978). The gate control theory of pain mechanisms. A re-examination and restatement. *Brain, 101*(1), 1–18.

Wallace, B., & Pagan, F. (2019). The implementation of overdose prevention sites as a novel and nimble response during an illegal drug overdose public health emergency. *International Journal of Drug Policy, 66*, 64–72

Wallace, B., Varcoe, C., Holmes, C., Moosa-Mitha, M., Moor, G., Hudspith, M., & Craig, K.D. (2021). Towards health equity for people experiencing chronic pain and social marginalization. *International Journal for Equity in Health, 20*(1), 53. https://doi.org/10.1186/s12939-021-01394-6

Wallis, C. (2019, April 19). Pain patients get relief from war on opioids. *Scientific American*. https://www.scientificamerican.com/article/pain-patients-get-relief-from-war-on-opioids1/

Weissman, D.E., & Haddox, J.D. (1989). Opioid pseudoaddiction - An iatrogenisyndrome. *Pain, 36*, 363–366.

Werb, D., Rowell, G., Guyatt, G., Kerr, T., Montaner, J., & Wood, E. (2011). Effect of Drug Law Enforcement on Drug Market Violence: A Systematic Review, *International Journal of Drug Policy, 22*(2), 87–94, https://pubmed.ncbi.nlm.nih.gov/21392957/#:~:text=Our%20findings%20suggest%20that%20increasing,markets%20can%20paradoxically%20increase%20violence.

West B., & Varacallo, M. (2022). *Good Samaritan laws*. Treasure Island, FL: StatPearls Publishing.https://www.ncbi.nlm.nih.gov/books/NBK542176/

What We Do. (n.d.). Retrieved from https://www.fda.gov/aboutfda/whatwedo/

Whitaker, B. (2015, November 1). Heroin in the heartland. *CBS News*. https://www.cbsnews.com/news/heroin-in-the-heartland-60-minutes/

White, J. (2001, February 10). VA police fear rise of new drug. *Washington Post*.

Whitehouse.gov. (2018, April 9). Fact sheets: President Donald J. Trump's administration is working every day to help bring an end to the opioid crisis. https://trumpwhitehouse.archives.gov/briefings-statements/president-donald-j-trumps-administration-working-every-day-help-bring-end-opioid-crisis/?utm_source=link&utm_medium=header

WHO. (1969). *Expert committee on drug dependence* (technical report series, no. 407, Geneva: WHO.

WHO. (1986). *Cancer pain relief*. Geneva: World Health Organization.

Williams, R. (1977). *Marxism and literature*. Oxford: Oxford University Press.

Wilson P.R., Caplan R.A., Connis R.T., Gilbert, HC., Grigsby, E.J., Haddox, J.D., Harvey, A.M., Korevaar, W.C., Cynwyd, B., Lubenow, T.R., & Simon, D.L., of the American Society of Anesthesiologists Task Force on Pain Management, Chronic Pain Section. (1997). Practice guidelines for chronic pain management. *Anesthesiology, 86*(4), 995–1004.

Wolford, B. (2021, August 6). Cop saved from fentanyl overdose in dramatic California video. "Not gonna let you die." *The Sacramento Bee*. https://www.sacbee.com/news/california/article253325223.html

Wootson Jr., C. (2017, June 24). A doctor prescribed so many painkillers, she's been charged with murdering her patients, authorities say. *Washington Post*. https://www.washingtonpost.com/news/to-your-health/wp/2017/06/24/a-doctor-prescribed-so-many-painkillers-shes-been-charged-with-murdering-her-patients-authorities-say/?utm_term=.1487fb6e2b37

Wydon, R. (2015, May 5). Letter to the Honorable Thomas E. Price. Secretary, United States Department of Health and Human Services. https://www.finance.senate.gov/imo/media/doc/050517%20Senator%20Wyden%20to%20Secretary%20Price%20re%20FDA%20Opioid%20Prescriber%20Working%20Group.pdf

Young, R.J., Mullins, P.M., & Bhattacharyya, N. (2021). The prevalence of chronic pain among adults in the United States. *Pain*. Advance online publication. https://doi.org/10.1097/j.pain.0000000000002291

Ziadni, M.S., You, D.S., Sturgeon, J.A., Mackey, S.C., & Darnall, B.D. (2020). Perceived injustice mediates the relationship between perceived childhood neglect and current function in patients with chronic pain: A preliminary pilot study. *Journal of Clinical Psychology in Medical Settings*. https://doi.org/10.1007/s10880-020-09722-8

Zigon, Jared. (2018). *A war on people: Drug user politics and a new ethics of community*. Berkeley, CA: UC Press.

Zola, I.K. (1972). Medicine as an institution of social control. *Sociological Review*, 20(4), 487–504.

Index

For the benefit of digital users, indexed terms that span two pages (e.g., 52–53) may, on occasion, appear on only one of those pages.

Figures are indicated by *f* following the page number

abuse-deterrent formulations, 59–62
   Purdue Pharma marketing and branding of, 92–93
acute pain, *versus* chronic pain, 30
addiction
   cultural representations of, 135
   psychological approach to, 134–35
   punitive approaches to, 134–35, 141–42, 143–45
   rerouting through problem of pain, 148–49
"America's New Drug Policy Landscape," 116
Angell, Marcia, 88
assessment of pain, 16–17
   evidence-based medicine and, 40–42
   pain as fifth vital sign, 21–22, 37–45
"awareness raising" campaigns, and marketing of OxyContin, 88–92

balloon effect, and abuse-deterrent formulations, 67–68, 149–50
Bandura, Albert, 107–8
Baszanger, Isabelle, 112
Beck, Ulrich, 54
Beecher, Henry, 24–25
behavioral psychology
   and multidisciplinary pain clinics, 27
Benet-Weiser, Sarah, 82–83
Biden, Joseph, 151
black-box warning, on OxyContin packages, 52–53
Bonica, John, 25–26, 26n.1, 28
branding practices, and opioid prescription drugs, 13–14
   branding of pain relief products, 76–79
   and consumer demand, 17–18
   future implications, 157
   governing clinical practice, 94–99
   governing opioid consumers, 79–94, 82*f*
   neoliberal capitalism and, 14
   and Purdue Pharma, 78–90
   "restoration" of self, 87–88
   self-help industry and, 18
branding practices, of Purdue Pharma, 78–90
   governing clinical practice, 94–99
   governing opioid consumers, 79–94, 82*f*
Burroughs, William, 132–33

Campbell, Nancy, 133–34, 140–41
cancer pain, morality of, 32–37
"Cancer Pain Relief," WHO report, 32–35, 34*f*
capitalism, and drug deregulation, 14
carfentanil, 71
Carpenter, Daniel, 48–49
Centers for Disease Control and Prevention (CDC), 137
characteristics of opioid epidemic, 6–7, 120
chronic-disease self-management program, 107–12
chronic pain
   biopolitics of self-help management, 115–17
   gaining legitimacy for, 136–40
   medicalization of, 29–31, 148–49
clinical practice, and branding pain relief, 94–99
commodity activism, and marketing of OxyContin, 93–94

Conquering Pain Act of 2001, 44–45
consumer demand, cultivating, 4
  and branding practices, 79–94, 82f
  and pharmaceutical branding, 17–18, 78–90
  "restoration" of self, 87–88
consumers, and users *versus* abusers, 69, 75, 75n.1
continuing medical education (CME), and marketing of OxyContin, 80–81
Courtwright, David, 132–33
COVID-19 pandemic, and harm-reduction services, 154–55
Crawford, Lester Mills, 64
criminal justice system, drug policy and, 151–53
criminal prosecution, patient deaths and, 36–37
cultural norms, and opioid epidemic, 147–48, 157

data collection, postmarket, 55–56
death
  criminal prosecution, patient deaths and, 36–37
  increased rates of, 6, 20
  opioids as cause of, 2–3
definition of opioids, 5–6, 5n.1
Derkatch, Colleen, 87–88, 105–7
Dole, Vincent, 134–35
drug markets, opioid epidemic and, 6–7
drug policy, and implications for the future, 149–59
Drug Policy Alliance, 153
Dumit, Joe, 87, 106–7

educational campaigns, and marketing of OxyContin, 80–81, 88–92
Egoscue, Pete, 103–4
emotional freedom technique (EFT), 104–5
Endo Pharmaceuticals, 61
evaluation of pain, statutory standard for, 39
evidence-based medicine, and pain management, 40–42, 153–55
expertise, institutional *versus* lived, 156

fentanyl
  challenges for regulators, 71–73
  and patterns of opioid use, 14, 69–73
  punitive measures against, 152
  and return to "war on drugs," 143–45
financial and legal costs, for Purdue Pharma, 76–78, 77n.2
Food, Drug, and Cosmetic Act of 1938, 48–49
Food and Drug Administration Authorization Act (FDAAA) of 2007, 54–55
Food and Drug Administration (FDA)
  and clinical trials for OxyContin, 48–50
  commissioners, ties to pharmaceutical industry, 64
  introduction of OxyContin to market, 46–48
  noninterventionist approach to opioid regulation, 62–66
  OxyContin and risk management framework, 50–53
  pharmacovigilance, 54–66
  postmarket data collection, 55–56
Food and Drug Administration Modernization Act (FDAMA) of 1997, 63–64
Fordyce, William, 27
future, implications for the, 149–59

gate control theory (GCT), 30–31
  origin of, 28, 28n.3
gender, and marketing of OxyContin, 92
genealogy, of pain management, 21, 22–23
geographic characteristics, of opioid epidemic, 14
Good Samaritan laws, 150, 152–53
Gottleib, Scott, 56–57, 64

Haddox, David, 98–99
harm-reduction policies, 144–45, 150, 153–55
Hayes, Arthur Hull, 64
healthcare disparities, and opioid treatment, 158
heroin, cultural representations of, 132–33
Hippocratic Oath, 23
Hoffman, Phillip Seymour, 121–22
Husel, William, 36–37

"I Got My Life Back: Patients in Pain Tell Their Story," 81–87, 82f
institutional expertise
  versus lived expertise, 156
  shortcomings of, 9–10
interventions
  and discourse of uncertainty, 8–10

moral interventions, 22–23
public health and criminal justice, 7
Intractable Pain Treatment Acts (IPTAs), 35–36
Inventing Pain Medicine (Baszanger), 112

Jenkins, John, 65–66
Joint Commission on Accreditation of Health Care Organizations (JCAHO), 41–42, 95–96
Junkie (Burroughs), 132–33

Kefouver-Harris Amendments of 1962, 48–49

legal costs, for Purdue Pharma, 76–78, 77n.2
lesion-free pain
  as diagnosis in its own right, 25–26
  and medicalization of chronic pain, 29–31
life expectancy, and opioid epidemic, 7
Living a Healthy Life with Chronic Pain (Lorig, Lefort et al.), 109–12
Lorig, Kate, 107–10

Management of Pain, The (Bonica), 25–26, 26n.1
markets, opioid prescription drug, 13
  and branding of pain relief products, 76–79
  "gray" market, 13, 66–74
McClellan, Mark, 64
McGee, Micki, 104, 106
McGill Pain Questionnaire, 28–29
McGoey, Linsey, 47
medical-industrial complex, and problematization of pain, 148–49
medicalization, of chronic pain, 29–31, 148–49
Meldrum, Marcia, 27, 32
Melzack, R., 30–31
modernization risks, 9
moralization of pain, 22–23
  morality of cancer pain, 32–37
morphine, nineteenth-century use of, 24
MS Contin, 50–51
Multidisciplinary Pain Clinic, University of Washington, 26

Naked Lunch (Burroughs), 132–33

naloxone
  access laws, 150
  introduction and use of, 140–43
neoliberal capitalism, and drug deregulation, 14
numerical rating scale (NRS)
  guidelines for clinical use of, 42–43
  introduction of, 28–29, 31
Nyswander, Marie, 134–35

objectivation of pain, 22–23
Opana, removal from market, 61–62
"Opioid Diaries, The," 1–3
opioid epidemic
  boundaries and categorizations, 13, 18–19
  and branding of pain relief products, 76–79
  characteristics of, 6–7, 120, 149
  characterizations of, 14
  cultural norms and, 147–48
  distinctions among people, 120–24
  interventions, public health and criminal justice, 7
  multifaceted phenomenon of, 19
  narratives connected to, 10, 18
  users *versus* abusers, 69, 75, 75n.1
  waves of, 69
opioid prescription drugs
  and evolution of pain management, 20–21
  genealogy of evolving use, 21, 22–23
  laws limiting prescription, 149–50
opioid prescription drugs, branding and regulation of, 13–14
  and "gray" market for opioids, 17, 66–74
  and self-help industry, 18
opioid regulation, strategic ignorance and, 73–74
  abuse-deterrent formulations, 59–62
  "balloon effect" in regulation, 67–68, 149–50
  and clinical trials for OxyContin, 48–50
  data collection, 55–56
  FDA noninterventionist approach, 62–66
  fentanyl, patterns of use, 69–73
  and introduction of OxyContin, 46–48
  knowledge gaps in pharmacovigilance, 54–66

opioid regulation, strategic ignorance and (*cont.*)
  and risk management frameworks, 50–53
  and risk of painkiller abuse, 53–54
opioids
  as cause of death, 2–3, 20
  cultivating consumer demand for, 4
  definition of, 5–6, 5n.1
Ortner, Nick, 104–5
OxyContin
  analysis of risks and benefits, 50–53
  clinical trials for, 48–50
  cultivating consumer demand for, 4, 17–18
  introduction to market, 46–48
  marketing and branding of, 79–88, 82*f*
  pharmacovigilance of, 54–66
  postmarket data collection, 55–56
OxyContin OP, 59–60

pain
  assessment of, 16–17, 21–22
  defining people in, 122–36
  as diagnosis in its own right, 25–26
  and discourse of uncertainty, 7–10
  as fifth vital sign, 21–22, 37–45, 100
  historical shifts in perspective, 22–23
  medicalization of chronic, 29–31
  networks of, 2
  parallel trends in opioids and, 3
  problematization of, 148–49
  problem of addressing, 3–7
  rhetorical-cultural analysis, 118–24, 145–46
  social construction of, 4–5
  subjective experience of, 155
pain assessment, 16–17
pain clinics, outpatient
  and "gray" market for opioids, 66–67
Pain Free (Egoscue), 103–4
painkillers, "gray" market for, 66–74
pain management
  do-it-yourself pain management, 103–7
  and evidence-based medicine, 40–42, 153–55
  and evolution of opioid use, 20–21
  genealogy of, 21, 22–23
  goal of universalizing, 28–29
  introduction of self-help to, 107–12
  patient-centered approach, 32–37
  pragmatic approach to, 23–24
  as universal human right, 16–17
pain medicine, introduction of field, 28–29
pain relief, and branding of products, 76–79, 100–1
  Purdue Pharma practices, 78–90
Partners Against Pain, 81–82, 89–90, 93–94
pathological consumer, 125–27
patient advocacy groups, and marketing of OxyContin, 81
patient-centered approach, to pain management, 32–37
patients
  gaining legitimacy for chronic pain, 136–40
  the pain-patient expert, 112–14
  pain patients-at-risk, 121–24
  pathological consumer, 125–28, 127*f*, 128*f*
  and self-help industry, 102
  and significance of self-reports, 31
  treatment-related deaths, 36–37
patient satisfaction, as metric for care, 96–98
patterns of opioid use
  fentanyl and, 14
pharmaceutical companies, ties to regulators, 157
pharmacovigilance, 8–9, 8–9n.2
  knowledge gaps in, 54–66
  as social technology, 54–55
physicians, criminal prosecution of, 36–37
"pill mill" operations, 66–67, 149–50
politics, and subjective experience of pain, 38–40
Portenoy, Russel, 139
prescription drug monitoring programs (PDMPs), 149–50
Prince, death from fentanyl overdose, 70–71
problematization of pain, 148–49
professional societies, and marketing of OxyContin, 81
pseudoaddiction, 98–99
Purdue Pharma
  branding practices, and cultivating clinical demand, 94–99
  branding practices, and cultivating consumer demand, 79–94, 82*f*

branding practices of, 78–90
and cultivating consumer demand for opioids, 4, 17–18
legal and financial costs, 76–78, 77n.2
OxyContin, introduction to market, 46–48
OxyContin OP, 59–60
and pharmacovigilance of OxyContin, 55–66
pharmacovigilance plan, 57–58

race, and characterizations of opioid epidemic, 7, 149, 152–53
regulation, of opioid prescription drugs, 13–14
analysis of risks and benefits, 50–53
"balloon effect," 67–68, 149–50
and clinical trials for OxyContin, 48–50
and "gray" market for opioids, 17, 66–74
and introduction of OxyContin, 46–48
neoliberal capitalism and, 14
and risk of abuse, 53–54
ties to pharmaceutical companies, 157
Researched Abuse, Diversion, and Addiction-Related Surveillance (RADARS) system, 57–58
rhetorical-cultural analysis of pain, 118–24, 145–46
Risk Society (Beck), 54

sales representatives, and marketing of OxyContin, 80
Scarry, Elaine, 82n.4
self-efficacy, and chronic-disease self-management, 107–12
self-help industry, and pharmaceutical branding, 18
self-help industry, and the pain-patient expert, 102, 112–14
do-it-yourself pain management, 103–7
introduction of self-help to pain management, 107–12
social class, and characterizations of opioid epidemic, 7, 149, 152–53
social construction of pain, 4–5
Social Security Disability Benefits Reform Act of 1984, 38–40

statutory standard, for evaluation of pain, 39
Stopping Overdoses of Fentanyl Analogues Act of 2019, 151
strategic ignorance, opioid regulation and, 73–74
abuse-deterrent formulations, 59–62
"balloon effect" in regulation, 67–68, 149–50
and clinical trials for OxyContin, 48–50
data collection, 55–56
FDA noninterventionist approach, 62–66
fentanyl, patterns of use, 69–73
and introduction of OxyContin, 46–48
knowledge gaps in pharmacovigilance, 54–66
and risk management frameworks, 50–53
and risk of painkiller abuse, 53–54
subjective experience of pain, 24–25
as the fifth vital sign, 21–22, 37–45
supply-side drug policies, 149–50

Tapping Solution (Ortner), 104–5
Team Against Opioid Abuse, 89, 90–92, 93–94
Trump, Donald, 144, 151
Tseng, Hsin-Ying "Lisa," 36

uncertainty, discourse of, 7–10, 11, 155–56
and introduction of OxyContin, 47
undertreatment of chronic pain
and morality of cancer pain, 32–37

Veterans Health Administration, and strategies for pain management, 41–42, 43–44
Visual Analogue Scale (VAS)
introduction of, 28–29, 31

Wall, P., 30–31
Weissman, David, 98–99
World Health Organization (WHO)
"Cancer Pain Relief" report, 32–35, 34f
pain ladder, 33–34, 34f